Enduring Geohazards in the Caribbean

Enduring Geohazards in the Caribbean

Moving from the Reactive to the Proactive

EDITED BY

Serwan M.J. Baban

University of the West Indies Press
Jamaica • Barbados • Trinidad and Tobago

University of the West Indies Press
7A Gibraltar Hall Road Mona
Kingston 7 Jamaica
www.uwipress.com

12 11 10 09 08 5 4 3 2 1

ISBN: 978-976-640-204-4

A catalogue record of this book is available from
the National Library of Jamaica.

Set in Sabon 10.5/14 x 27

Book and cover design by Robert Harris.

Printed in the United States of America.

To my family, Daya Dora, Judith, Shereen and Zana,

for sharing their son, husband and father with the rigours of academia.

Thank you for your love, support and encouragement.

Contents

Preface

The states in the Caribbean have a number of common characteristics that make them vulnerable to geohazards. These include geography, climate/weather conditions, limited physical size, finite natural resources, dependence on agriculture, tourism, and high population densities concentrated in vulnerable areas, that is, hillsides and flood plains. In addition, the region is experiencing rapid economic development combined with a fast rate of urbanization, population growth and questionable agriculture practices. These factors typically lead to floods, landslides, deforestation, soil erosion, and extinction of an unknown number of animal and plant species.

The economic, environmental and social costs of annual flood and landslide events amount to millions of dollars in the Caribbean region. Additional costs in terms of disruptions to the social fabric, damage to the flow of goods and services (for example, lower output from damaged factories, lost productivity and so on), and short- and long-term impacts on the environment and economy remain non-quantified.

Geohazards research in the region has been selective, project based, intermittent and sporadic, which does not lend itself to holistic understanding. A more rigorous approach is needed to enable scientific co-ordination and agreement and to allow for conclusive management approaches to emerge and be implemented.

The current management of floods and landslides is subjective and reactive as the major effort remains in cleaning-up-operations post event. Mitigation works are designed to repair infrastructure after the event has occurred. Clearly, there is an urgent need for objective decision making and for moving geohazards management from being reactive to proactive. However, the lack of an effective and reliable information base makes this transformation difficult. For example, at present there is an absence of a national data depository for hazard events, where event occurrences can be recorded and quantified for post analysis.

Nevertheless, there are clear indications that the information poverty obstacle can be managed by using reputable technologies that facilitate management decisions, such as geoinformatics, which encompass remote sensing, geographic information systems (GIS) and global positioning systems (GPS). Geoinformatics contains the necessary tools to collect, handle and analyse the necessary data sets, as well as to expand our knowledge of the processes involved at the appropriate scales. Furthermore, several governmental agencies seem to be responsible for geohazard management. These agencies are not capable of handling geohazards on their own, nor is there effective coordination between them.

The objective of this book is to contribute, in a small way, to promoting awareness among academics, geohazards specialists, users and policymakers, of the nature and extent of geohazards-associated problems and of the range of possible solutions to manage floods and landslides in a sustainable fashion.

These objectives are being addressed through:

1. Developing and promoting the holistic approach for managing geohazards in the region.
2. Providing a conceptual framework for transforming geohazards management from reactive to proactive mode.
3. Providing, demonstrating and evaluating the use of available and reliable cutting-edge technologies, such as GIS, remote sensing and GPS for managing geohazards.
4. Developing and demonstrating the use of national-level geohazards inventories and databases; early warning systems; predictive understanding of landslides and floods processes and triggering mechanisms; building resilient communities; and setting international-level standards for all consultancies.
5. Promoting effective programmes for public awareness, education and information, as well as enhancing the implementation capabilities of relevant government agencies.

Serwan M.J. Baban

Acknowledgements

This book owes its origin to the regional workshop Enduring Geohazards (Landslides and Floods) in the Caribbean Region, held in the Learning Resource Centre at the University of the West Indies, St Augustine, Trinidad, on 8 December 2004. The workshop was organized by the Centre for Caribbean Land and Environmental Appraisal Research (CLEAR) at the University of the West Indies and was supported by the British Council Higher Education Scheme and the Office of Research at the University of the West Indies. The workshop brought together experts from the Caribbean region to discuss geohazard issues and problems, and to intensify efforts towards a coordinated approach to manage them.

The workshop identified a number of strategies to handle geohazards in the region. Among them were the need to develop holistic and scientifically based management approaches, identify and map critical slopes using early warning systems, as well as use new technologies such as remote sensing and geographic information systems.

The book, which has been supported by the RBTT bank in Trinidad and the Office of Research of the University of the West Indies, results from the meeting on the university's St Augustine campus as well as invited contributions made by established geohazards management, development and planning experts from the Caribbean and worldwide.

I would like to thank my colleagues for contributing to the book and, in particular, for tolerating my reminders and for responding positively on most occasions. My thanks also to Greg Luker, the GIS lab manager at Southern Cross University, for assisting with the illustrations.

CHAPTER 1

Enduring Landslides and Floods in the Caribbean Region

ANGELLA CROPPER

Abstract

This chapter will explore the vulnerability of the Caribbean region to geohazards by looking at the nature of its exposure. It will argue that there is little that the region can do to avoid geohazards that ensue from its geographic and geologic situation, or are impacts of global forces over which the region has no control. However, given the vulnerability of the region to these events, as reflected in their incidence and scale, and in their associated dislocations and direct and indirect costs, the region could better manage its vulnerability to the effects of such events by building its resilience, preparedness and adaptation. The chapter proposes some approaches to this and suggests that these may be the most fruitful areas for intervention in preparing the region for "enduring" geohazards. The chapter draws conceptually upon the work of the Millennium Ecosystem Assessment and empirically on the findings of an assessment of the Northern Range of the island of Trinidad, Trinidad and Tobago.

1.1 Introduction

Geohazards can include the range of geological, ecological or hydrological processes or events which cause, or have the potential to cause, widespread damage to the environment and physical property, often involving injury and death to people in affected areas. Some are unpredictable and unavoidable. Others ensue as a result of the ways in which human activity alters or affects the functioning of natural systems (Cropper 2004).

Among them are events that have rapid onset, for which there may not be any appreciable warning or information. Such rapid onset events include earthquakes, volcanic eruptions, hurricanes and floods. Among them are also events, which have long gestation, for which symptoms can usually be seen, and which may be due to either natural or human causes, or combinations of these. These cover a range of processes, which may become hazardous only at later stages of their development, when underlying and continuous processes manifest themselves in hazardous occurrences. These would include events such as soil erosion, landslides and subsidence; sea level change and salt intrusion; deforestation and flooding; salinization, desertification and dust storms; siltation; simplification of landscapes; and reduction in biological diversity. Altogether, such processes generate loss of productivity of the natural resource base, which could lead to additional vulnerability to livelihoods, human health problems, and general loss of well-being.

Over the past decade, the world has experienced a spate of such "natural" disasters affecting about 2.5 billion people, killing close to 500,000 and causing economic loss estimated at US$700 billion. The risks of such hazards and people's vulnerability to their effects are for the most part not preventable, but there is a great deal that can be done to make us less abject in simply "enduring" such events and to enable us to manage our vulnerability.

1.1.1 Exploring Vulnerability

The Millennium Ecosystem Assessment (2003) defines "vulnerability" as "the capacity to be wounded by socioeconomic and ecological change". In practice, a close correlation is observed between changes in ecological services (decline in the benefits which humans receive from well-functioning ecosystems), and negative consequences for groups and individuals of low socioeconomic status. Vulnerability is therefore viewed as the product of interaction between environmental factors and socioeconomic and political systems.

Vulnerability is also a "measure" of the combined and interrelated impact on persons, groups or places of the exposure to geohazards (derived from factors outside the creation or control of people), their sensitivity to impacts of such events (depending on the nature and extent of the links between their well-being and ecosystems), and the *resilience* of people who are impacted by such events (based on levels of awareness, preparedness, and capacities to insulate themselves, respond or recover from impacts).

1.2 Exposure of the Caribbean to Geohazards

Exposure of the Caribbean to geohazards is derived from four major sources (Cropper 2004):

1. Geography: Most of the Caribbean is located within the tropics. This geographical location, which renders it susceptible to rainstorms and hurricanes, when combined with some of its geological features, exacerbates impacts through flooding and landslides. In addition, situated along the rim of the Caribbean Plate, the region is susceptible to earthquakes and volcanic eruptions.
2. Geology: In the volcanic islands of the Eastern Caribbean, with steep slopes prone to erosion, as well as lithological characteristics (for example, in Jamaica, Puerto Rico, and Trinidad and Tobago) there is increased exposure to geohazards from landslides and erosion.
3. Climate change: Changes in weather patterns – drier and wetter seasons interspersed with more extensive dry or wet periods – have been observed in the Caribbean over the past decade. There now seems to be a movement towards global scientific consensus that these observations, in the Caribbean and around the globe, are indications of long-term trends towards climate change. The exposure of the Caribbean to anticipated impacts of climate change – land-ocean interactions through sea level rise and salt water intrusion;

changes in frequency and intensity of storms; damage to coral reefs and coastal strips including infrastructure facilities; and unpredictable changes in weather patterns – is well rehearsed in the climate change impacts literature and inscribed in many conclusions of intergovernmental processes. Notable among these are the United Nations Conference on Environment and Developments (1992); United Nations Conferences on Small Island Developing States (SIDS) (1994) with its Programme of Action for SIDS; World Summit on Sustainable Development (2002); and Mauritius Conference on SIDS (2005).

4. Changes in ecosystems: The combined effects of geography, geology and climate change, together with human dependence on and the use of natural assets, and human impacts on the environment, increase the exposure of the Caribbean to another source of geohazards – degradation of its environmental parameters and decline in the capacity of its ecosystems to generate benefits ("ecosystem services"). The Millennium Ecosystem Assessment (2003) classifies benefits of ecosystems to humans ("ecosystem services") into provisioning (food, water, fibre, fuel), regulating (climate, water and disease regulation), supporting (primary production and soil formation) and cultural (spiritual, aesthetic, recreation, education) services (Cropper 2004). The well-being of all societies is dependent on these services – in different mixes, with different manifestations. The Caribbean is heavily dependent on the ecosystem base for its well-being: many of its economies rely predominantly on agriculture, fisheries and nature tourism; most of the population live within coastal zones; and natural and human causes of loss of mangroves, damage to coral reefs, degradation of coastal strips, loss of forest cover, and pollution of surface and groundwater are many and widespread.

1.2.1 Sensitivity of Human Well-being to Exposure to Geohazards

The relationship of human well-being to such ecosystem changes determines the "sensitivity" of any group or place to exposure to geohazards. The Millennium Ecosystem Assessment (Figure 1.1) illustrates the determinants and constituents of human well-being as the following:

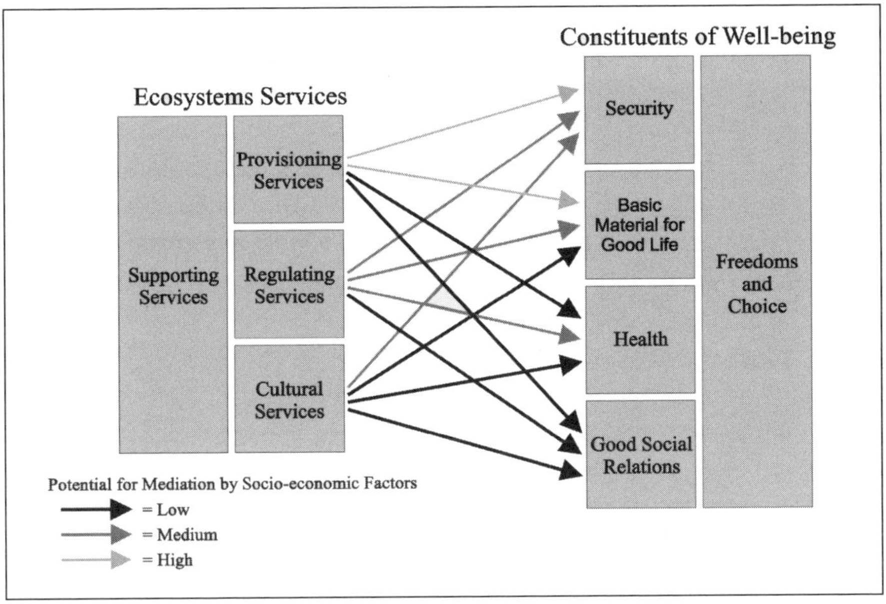

Figure 1.1 Relationship between ecosystem services and human well-being (Millennium Ecosystem Assessment 2003).

1. Security: ability to live in an environmentally clean and safe shelter; ability to reduce vulnerability to ecological shocks and stresses.
2. Basic material for a good life: ability to access resources to earn income and gain a livelihood.
3. Health: ability to be adequately nourished; ability to be free from avoidable disease; ability to have adequate and clean drinking water; ability to have clean air; ability to have energy to keep warm and cool.
4. Good social relations: opportunity to express aesthetic and recreational values associated with ecosystems; opportunity to express cultural and spiritual values associated with ecosystems; opportunity to observe, study and learn about ecosystems.
5. Freedoms and choice: ability to realize one's potential and capacities and the opportunity and means to do so.

The degree of sensitivity will depend on the nature of the sources of exposure and the extent to which these constituents and determinants of well-being are reliant on or related to the natural world. Figure 1.1 is a conceptual illustration of that relationship. It will be readily appreciated that any "measure" of sensitivity for any group or place will depend on an amalgam of factors, spanning the range of natural, demographic, economic, governance, institutional, and cultural considerations. Moreover, the degree of "sensitivity" would also include values that may be dominant or absent within the group or society, and estimating the strength of the relationship between ecosystems and human well-being would involve some degree of subjectivity (Cropper 2004).

1.3 How "Sensitive" Is the Caribbean to Geohazards?

Caribbean sensitivity to geohazards can be gleaned by citing some examples of such historical events in the Caribbean, characterizing the nature and extent of the impacts, and noting the level and distribution of costs, direct and indirect.

In Venezuela, in 1999, 350,000 people in Vargas State were affected, including 30,000 who died, from mudslides on the hillside settlement of very poor communities; 200,000 were left homeless; damage was estimated in billions of dollars; and reconstruction, relocation and resettlement would require several years of effort.

In Honduras, in 1998, sudden flooding left 6,500 people dead; 11,000 missing who were presumed dead; 1.5 million (about 20% of the population) homeless; between 70% and 80% of the transportation infrastructure destroyed; 70% of crops destroyed (valued at US$900,000); food, water and medicine shortages; and episodes of malaria, dengue and cholera.

In the Eastern Caribbean, over the last 300 years, it is estimated that 30,000 have perished from volcanic eruptions, 15,000 from earthquakes, 15,000 from hurricanes and 50 from tsunamis. Additionally, for the Caribbean, the average number of deaths from hurricanes per year over the period 1980 to 2000 has increased from 10 to 200.

In Haiti, over the last two decades, loss due to flooding is estimated as US$5 billion. In 2004, Hurricane Jeanne yielded floods that caused the deaths of 2,700 people.

In Grenada, in 2004, most infrastructure, including electricity and communication systems, as well as 90% of all buildings, suffered structural damage from Hurricane Ivan.

In Montserrat, a volcanic eruption in 1995 caused the complete destruction of the capital city, Plymouth, and the entire economy of the island. The damage is estimated at US$500 million.

From such events, the direct costs in terms of human life and health, damage to property, cost of clean-up and repairs of infrastructure and housing, and long-term economic costs of rebuilding, compensation, and the like, can readily be estimated. Who bears these costs? Poor groups are the most seriously and directly affected, because of their precarious locations, poor housing, no insurance cover, land-based livelihoods and few options. There is a very close link between general level of development in a society and degree of vulnerability (UNDP/UNEP 2004). The societies in general bear the immediate and extended costs of rescue, rehabilitation and recovery. Occasionally, the costs accrue beyond their borders: foreign aid to Haiti, for example, amounted to US$500 million in 2004.

But the indirect costs can be even more significant and long lasting. Social costs include loss of whole communities and towns, disruption in culture and traditions, ongoing trauma from loss of loved ones and from the experience, displacement of families, and demoralization of governments and societies. Economic costs include losses of or radical changes to livelihoods, and loss of productive assets like topsoil. Environmental costs include loss of productivity; unplanned and unmanageable settlements as displaced people seek new locations for dwellings, involving deforestation and land conversion for agriculture; and further inadequate infrastructure and services.

The cumulative and long-term effects of these consequences of geohazards are far reaching, including rendering the afflicted even more exposed and more vulnerable in another round of such events.

1.4 The Northern Range of Trinidad, Trinidad and Tobago

The sensitivity of Trinidad and Tobago can be imagined based on the findings of the Northern Range Assessment (2005). The Northern Range is a continuation of the Coastal Cordillera of Venezuela, stretching across the northernmost quarter of the island of Trinidad, with contours generally between 90 m and 450 m, but with some elevations of over 600 m. It is rugged topography with steep slopes, more than 80%

of which have a gradient of 20° or higher. Its geological structure and soils combine to render the hillsides prone to soil erosion and land slippage.

The assessment reveals that significant land-use changes – unauthorized human settlements, market agriculture on steep slopes and on lands not classified for agriculture, and agricultural lands predominantly being used for housing, change in forest cover, and quarrying – have taken place, with the pattern of use moving eastwards and upwards into the valleys. This pattern is accompanied by, and may even be caused by, little societal understanding of how we affect our natural systems and the consequences for our well-being, as well as inadequate planning and inadequate enforcement of policies.

The sources of exposure to geohazards described earlier apply to Trinidad and Tobago as they do to the rest of the Caribbean. The experience of Vargas State in Venezuela, cited earlier, in which houses and people slid down the hillsides could occur in Trinidad and Tobago, with the difference being only in scale. The experience of South and South East Asia with the tsunami of December 2004 could occur in the Caribbean given its geographical location at the convergence of the Caribbean and South American tectonic plates.

1.5 Conclusions and Recommendations

The Caribbean is not able to avoid the potential for geohazards to which it may be exposed because of its geography and geology. It can do little to alter the course of climate change, although it can take measures to adapt to the impacts of the process of climate change. It can avoid to some extent drastic changes in ecosystem capacity to continue to provide regulating and supporting services, but it can manage better its vulnerability.

Mitigating its circumstances will depend on how it builds *resilience*, how it establishes preparedness, and how it organizes for adaptation to climate changes that appear to be taking place. Resilience is defined as the amount of disturbance a natural system can absorb while maintaining basic functions, or the degree to which a social system is capable of self-organization and building its capacity for learning and responses.

All of these are facets of the response mechanisms that can be addressed concurrently through approaches and measures of which Caribbean societies are capable, and will reduce the human devastation that accompanies hazardous events.

Increasing the ability of Caribbean societies to manage vulnerability will require focused attention to the following approaches and measures if the Caribbean is to transition from reacting to geohazards to preparing to manage its vulnerability to them:

1. public policy that reflects understanding of sources of exposure and vulnerability, and consciously addresses how human contributions to causal factors and human consequences from events can be minimized;
2. creation of a scientific information base and the carrying out of assessments and vulnerability mapping;
3. development planning that is based on scientific assessment, geohazards and vulnerability mapping and utilizes a preventative approach to degradation and risk exposure, including zoning according to land capability and regulating settlements, infrastructure, and building; spatial and urban planning that recognizes sources of exposure and nature of risk;
4. preparedness through enhancing monitoring and early warning systems (such as the Caribbean Community Climate Change Centre, or the Seismic Research Unit of the University of the West Indies); making national and regional response mechanisms (such as the Trinidad and Tobago National Emergency Management Agency or the Caribbean Community Emergency Disaster Response Agency) effective and efficient; and ensuring technical preparedness in the use of sensing technology, information and communication systems;
5. adaptation through all means possible to the anticipated impacts of climate change – among them community-based sustainable livelihoods, reforestation, appropriate building design and codes, physical planning, conserving mangroves, seagrass beds, and rehabilitating coral reefs;
6. affecting culture and behaviour through public awareness and education to reduce complacency and convey to the public that managing vulnerability is everyone's responsibility; enabling societies to

discharge that responsibility, making use of incentives and penalties as appropriate; and

7. regulation, implementation and enforcement of policy approaches, and other measures that are undertaken.

Many of these elements exist in varying degrees throughout the Caribbean. (The degree to which they exist and are ingrained in the response mechanisms of a society might explain the differential effects of Hurricane Ivan as between Grenada and Cuba, or of Hurricane Jeanne as between Haiti and the Dominican Republic.) But they are, in general, rudimentary, disparate and uncoordinated. Often there is no mechanism – conceptual, policy or operational – that brings them together to understand better the vulnerability, to be efficient and effective, and to have the whole exceed the sum of parts. So there is a need to put these elements together and to fill the gaps, in order to better organize to manage our vulnerability.

A useful starting point would be to build a robust conceptual and planning framework for understanding and assessing risk, for linking to human well-being and for identifying points of resilience. Such a framework would seek to

1. clarify the natural and human driving forces of vulnerability;
2. demonstrate the relationship between human well-being and integrity of ecosystems;
3. illustrate the relationship between poverty and vulnerability;
4. track stresses and perturbations to better understand cumulative impact;
5. indicate the role of organizations and technical programmes in mitigating risk and managing vulnerability;
6. improve the knowledge base of patterns of vulnerability;
7. improve assessment methods and tools and build capacity for risk management; and
8. identify the scope and nature of interventions that would reduce risk and vulnerability.

All of the above presumes recognition of the importance of the precautionary principle and its extension to safeguarding human life, human health, the economy and the environment.

Acknowledgments

The author acknowledges the contributions to the paper of her colleagues in the Cropper Foundation, Keisha Garcia and Sarika Maharaj.

References

Cropper, A. 2004. Enduring geohazards in the Caribbean region. Paper presented at the workshop Enduring Geohazards (Landslides and Floods) in the Caribbean Region, Learning Resource Centre, University of the West Indies, St Augustine, Trinidad. 8 December.

Mauritius Conference on SIDS. 2005. http://www.un.org/events.

Millennium Ecosystem Assessment. 2003. *Ecosystems and human well-being: A framework for assessment.* Washington, DC: MA and Island Press.

Northern Range Assessment. 2005. *Report of an assessment of the Northern Range, Trinidad, Trinidad and Tobago: People and the Northern Range.* State of the Environment Report 2004. Port of Spain: Environmental Management Authority of Trinidad and Tobago.

United Nations Conferences on Environment and Development. 1992. http://www.un.org.

United Nations Conferences on Small Island Developing States. 1994. http://www.un.org.

UNDP/UNEP 2004. Reducing disaster risk: A challenge for development. Report. http://www.un.org.

United Nations Environment Programme (UNEP). 2005. Report of the Global International Waters Assessment. http://www.un.org.

United Nations International Strategy for Disaster Reduction (UNISDR). 2006. http://www.unisdr.org.

World Summit on Sustainable Development. 2002. http://www.un.org/events.

SECTION 1

Landslides

CHAPTER 2

Modelling Landslides in Tropical Environments

KEITH TOVEY

Abstract

Several methods exist for analysing and managing the consequences of landslide hazards. These range from the purely engineering approach, in which detailed analysis of selected slopes can be done to assess the likelihood of failure of those specific slopes, to the analyses based on a geographic information systems (GIS) approach, which explore the previous incidence of landslides and can relate landslide locations to the prevailing geology, soil type and land use/cover type in an area. Landslides can cause not only loss of life, but they also disrupt the economic activity of a region. Steps must be taken to ensure that such losses are minimized, and a proactive approach to landslide hazard management is needed. Such an approach requires that a rational database of areas prone to landslides is developed, and this in turn requires that a cost-effective method is available to capture the initial condition of slopes.

In tropical countries, the manifestation of landslide hazards is often associated with roads, and this provides an effective method to capture the required data needed to categorize areas either prone to landslides or areas which appear to be free of them. The development of such a technique is described in this chapter, using examples researched in Trinidad.

2.1 Introduction

The true hazards posed by landslides are often masked within other geological hazards such as earthquakes or extreme climatic events, for example, hurricanes. Even the largest landslides are of limited geographic extent, and the economic and social impacts of landslides are often not clearly recognizable as they are considered to be merely a part of the major climatic or earthquake event (Ahmad and McCalpin 1999). In Hong Kong in 1972, for instance, a disastrous rainstorm caused two major landslides resulting in the deaths of over 140 people, and yet the official report refers to them as the "Rainfall Disasters of June 18th 1972" (Schoustra 1972).

Landslides occur when the disturbing forces exceed the resisting forces in the soil mass. These resisting forces are closely related to the shear strength of the in situ soils and any associated pore water pressure. Landslides often occur on hillsides unaffected by human activities and have been instrumental in the formation of the present morphology: some of these are large, such as the Mam Tor landslide in Derbyshire, England (Skempton et al. 1989), and many on the south side of the Northern Range in Trinidad. However, while such landslides do still occur in relatively uninhabited regions, many of the landslides occurring at the present day, and which directly affect the local population, are partly caused by anthropogenic action.

Four fundamentally different types of slope type are shown in Figure 2.1. These four types may be defined as follows:

1. *Cut slopes:* Slopes on which the natural, geologically evolved slope has been steepened by human activity to provide a level area for a building or a road. A cut slope will affect the lower part of a slope and may have above it the unmodified slope profile (Figure 2.1a). The act of creating the cut slope will modify the failure mechanism, which may be extensive and may potentially cause a landslide including the "natural" unmodified slope above; for example, the

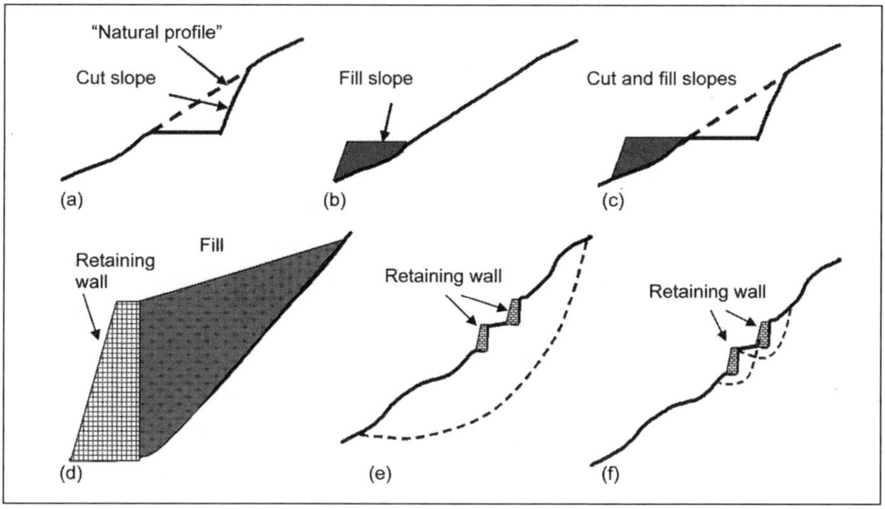

Figure 2.1 Types of slopes and failures: (a) cut slope, (b) fill slope, (c) cut slope above a fill slope to provide a wide platform for road building, (d) retaining wall, (e) general failure in an extensive slope: the presence of terracing has little effect on stability, (f) localized failures of terraces would be classified as a retaining wall failures.

Po Shan Road landslide in Hong Kong in 1972 (Schoustra 1972; GEO 1992; Cruden and Varnes 1996).

2. *Fill slopes:* These are slopes that have been created by placing excavated material onto the unmodified slope profile (Figure 2.1b). The purpose of this is to extend a flat area and may involve material placed over an existing slope to steepen it. In road construction, it is common to find the platform created by forming a cut slope on one side and a fill slope on the other (Figure 2.1c).
3. *Retaining walls*: These are not really slopes, but they play an important role in the stability of several slopes. A retaining wall may be used to retain a level fill area, or alternatively it may be used in conjunction with a cut or fill slope to improve the stability of the latter (Figure 2.1d). In most cases, the retaining wall is located at the base of the cut or fill slope. However, in Trinidad, there are many instances in the Central Range of hills where retaining walls are constructed on top of fill slopes, sometimes in an attempt to reconstruct a road built on the unconsolidated fill material.

4. *Geometrically unmodified or "natural" slopes*: These are slopes on which anthropogenic activity is (or has been) of limited extent. A "natural" slope will be one in which human activity has not caused any change in the primary mode of failure (that is, not covered by cut or fill slopes). For example, such a slope will exist in places where vegetation has been changed by man, thus affecting the run off characteristics. The resulting changes in the water table have a secondary effect on the mechanism of failure, which is unlike the substantive change in the slope profile associated with a cut or fill slope. A large slope on which terracing has taken place with small retaining walls of 1 m to 2 m height would still be classified as "natural" (Figure 2.1e) and the general stability of the slope will be dictated by a slip circle, which is modified only in a minor way by the presence of terracing. On the other hand, the failure of an individual retaining wall on terraced slopes (Figure 2.1f) would not be considered as a "natural" failure, since at the scale of this local failure, the anthropogenic activity would have been the *primary* cause of failure.

The primary causes of landslides are numerous, and in some cases, unexpected causes have been identified. For example, trees on a slope are often seen as an effective means of stabilization, as they not only provide soil reinforcement via their roots, but also help to reduce the local ground water table. However, those species which have deep tap roots can be detrimental, since in windy conditions the movement of the trees can cause voids around the roots, which then allows easy ingress of water, thereby increasing the pore water pressure. The present author observed several such failures while carrying out landslide emergency duties in Hong Kong in 1982. However, in most cases, though it is a combination of effects that cause a landslide, it is only one of these effects that finally triggers the landslide to occur at a particular time and place.

The pressure to find suitable land for buildings and highways has increased the anthropogenic modification of slope profiles. This in turn has increased the risk of landslide hazards, particularly in tropical and semi-tropical countries where significant interruption to lines of communication and death or injury can occur. Often, there has been a reactive approach to dealing with such hazards, with hastily prepared remedial works, which are definitely not the most effective in the long run. A move towards a more proactive approach is essential, but there will always be a conflict over resources. This chapter examines how these resources may be used effectively in tropical countries by using data collected in Trinidad and Tobago and by drawing on examples of the pioneering work done in places like Hong Kong in the late 1970s and early 1980s.

2.2 Geography and Geology of Trinidad

The Republic of Trinidad and Tobago has the total area of some 5,128 km^2, located about 12 km off the north coast of Venezuela on the South American mainland and lying between 10° and 11° north (Figure 2.2). Trinidad accounts for 94% of the total area and about 96% of the total population, which was estimated to be 1.09 million in 2005.

The climate of Trinidad and Tobago is tropical, with an average annual temperature of approximately 27°C but with diurnal temperature variations of the order of 8°C. The average annual rainfall for Trinidad is 1,869 mm, most of which occurs in the wet season between June and December. The highest rainfall is recorded in the Northern Range where there can be as much as 3,200 mm per annum. In Trinidad, there are three mountain ranges. In order of size they are the Northern Range (up to 940 m), the Central Range (up to 336 m) and the Southern Range (up to 330 m). While, at present, the majority of landslide failures affect roads, there is increasing pressure to develop areas in the steep Northern Range. Failures affecting other developments are likely in such areas in the future if careful management procedures are not adopted. The types of landslide failure are very different in the three ranges, reflecting the different geology and terrain of each one. Whereas the majority of slopes in the Northern Range are of the cut slope type, those in the other ranges are usually of the fill type and are sometimes associated with the failure of a retaining wall. Failures on cut slopes may block a highway for a period of hours to days, but large failures in fill slopes often result in the complete destruction of the road that may take months to be reinstated.

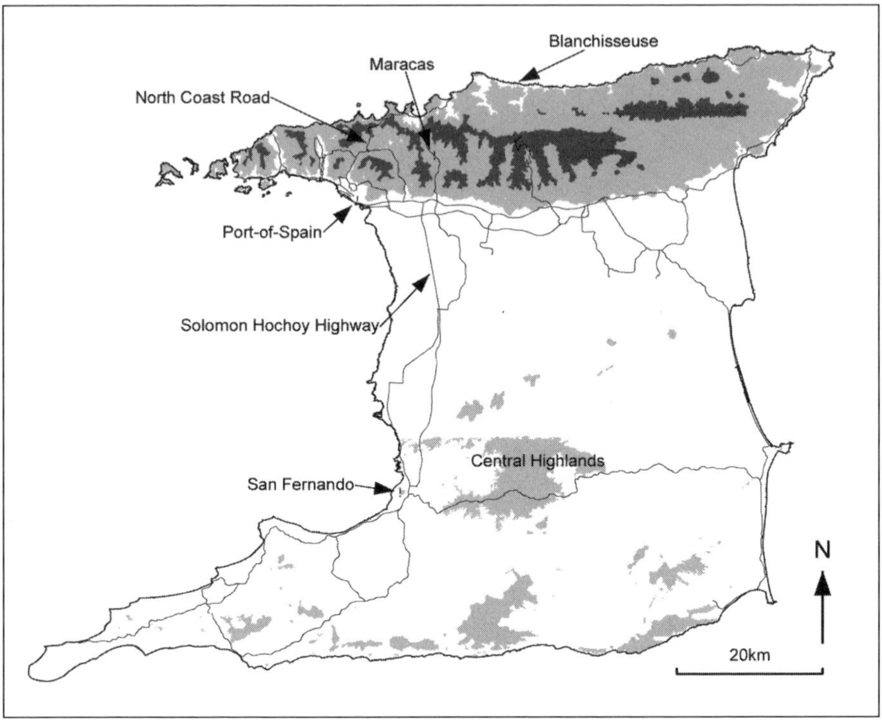

Figure 2.2 Map of Trinidad showing main mountain ranges and roads referred to in the text.

Any classification of slopes or potential landslide hazards will need to differentiate between these different slope types and the associated modes of failure. Over 95% of landslides occur in the months of August to December (Figure 2.3). However, despite the high incidence of landslides occuring during the wetter part of the year, there has been no systematic recording of occurrence and the associated rainfall at the time. This makes it difficult to develop a reliable warning system such as that which is presently employed in Hong Kong. The peak occurrence in August is perhaps unexpected, as the month with the highest rainfall is often November. However, this peak occurrence probably arises from sub-aerial weathering, the cutting of new slopes and the fact that the prolonged preceding dry spell would have contributed to negative pore pressures in the existing slopes. The onset of the first significant rainfall weakens the slopes, causing the high incidence of landslides in August.

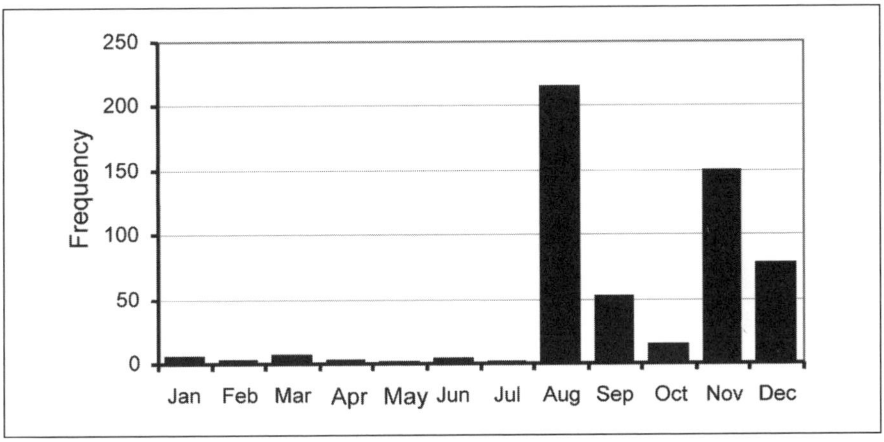

Figure 2.3 Landslide occurrences throughout the year. Derived from data provided by Gay (2004).

The angle of the slopes on which landslides have occurred is very different between the Northern and Central Ranges (Figure 2.4), reflecting the differences in the slope types in the two ranges. Most of the recent failures in the Northern Range are cut slope failures, associated with the main highways, particularly the North Coast Road, and platforms recently cut for development. Many of the failures affect not only the cut slope itself but also the "natural" slope above. Occasionally there are indications of the "natural" slope above a cut slope failing, while

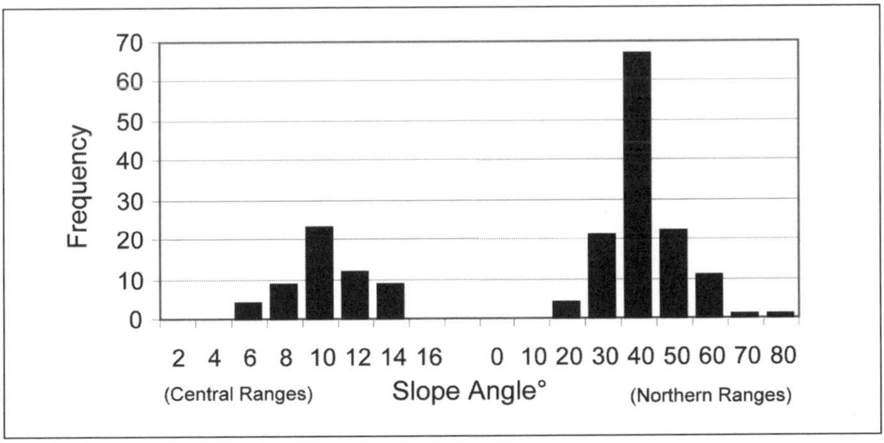

Figure 2.4 Landslide frequency on slopes in Central and Northern Ranges (Gay 2004).

the cut slope below remains undisturbed. An example of this occurred at chainage 14+100 km (Easting 672189 Northing 1189719) on the North Coast Road during the rainstorm of 19 November 2003. This was a major slide above an intact 30 m high cut slope and was 50 m wide. The debris blocked the road for about 18 hours.

A few extensive fill slope failures on the North Coast Road also occur. These sometimes result in debris flows up to 100 m to 200 m in length and are not infrequently associated with a nearby major cut slope failure, which block the road, causing a diversion of the drainage flow over the top of the fill slope where the failure then occurs. Such an example occurred on 9 December 2004 at chainage 23+000 (Easting 678201 Northing 1193705) and extended at least 80 m down slope. This was associated with a major cut slope failure that occurred on the same date on the opposite side of the road.

There is evidence to suggest that rainfall alone may not be the sole cause of landslides in fill slopes in the central highlands, as some have occurred where there have been leakages from water mains – for example in November 2003 at approximately chainage 5+300 km on the Indian Trail Road in Central Trinidad. Even carefully engineered fill slopes have not escaped failure, as was evident on the embankments to the flyover across the Solomon Hochoy Highway at the Claxton Bay Interchange in November 2003, and the Indian Trail overpass in December 2004.

2.3 Analysis Methods

There are three basically different approaches to landslide analysis, all three of which may be incorporated as proactive management and planning tools: (1) an engineering approach, (2) a GIS-based approach and (3) a landslide warning system approach.

1. An engineering approach

The first approach involves a traditional engineering approach, which is deterministic and involves detailed numeric modelling and analysis of the slope. It can be costly to undertake and is data-intensive, requiring and the detailed measurement of the slope profile soil properties, as well as an accurate definition of the location of the water table. Such analyses may be used to ensure that a minimum factor of safety (F_s) for a slope is reached, and they are essential for new developments. The engineering approach may be summarized in Figure 2.5. Central to this approach is the stability assessment which will be drawn from key physical factors such as the slope profile, both the ground water and surface hydrology, any load on the slope, and finally, but perhaps the most important, the inherent material properties of the soil. The factors are in turn influenced by anthropogenic activity and the underlying geology and soil types.

When a landslide occurs, there are essentially two options available to deal with the consequence. The first is to remove the consequence, and the second is to initiate remedial works. Good examples of the first option are the removal of squatter huts from areas affected by landslides in Hong Kong and the abandonment of the A625 main road over Mam Tor in Derbyshire, England in the early 1980s.

Removal of the consequence should be followed by stability assessments, as the profile and hydrology of the slope will have been modified

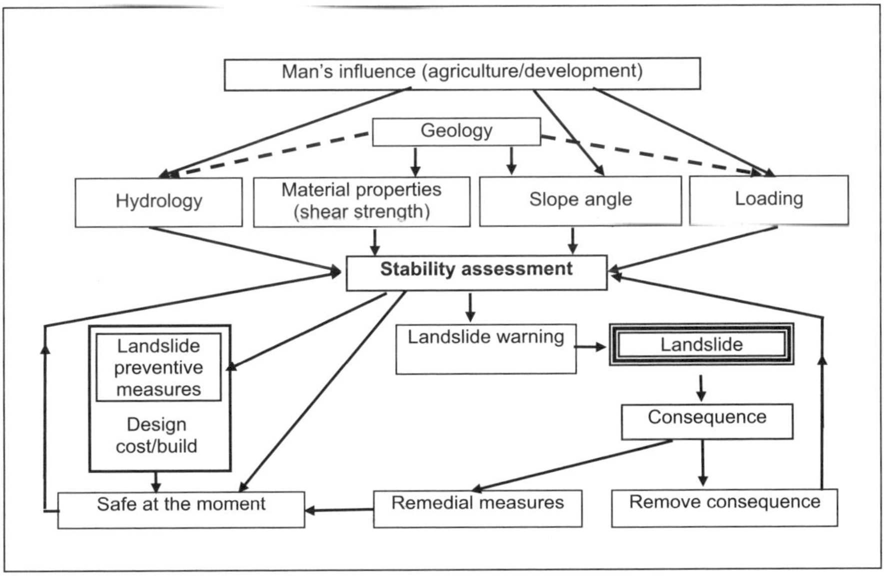

Figure 2.5 The engineering approach to landslide assessment and management.

by the landslide, and future landslides may affect the same area. Remedial works will make the slope at least temporarily safe, but once again stability assessments should be undertaken to assess the long-term stability.

A slope that is just stable will have a factor of safety, as unity, while the more stable a slope is, the higher the value of F_S. While, in theory, there should be no existing slopes with a factor of safety less than unity, it is not uncommon to find these in the field, and their presence reflects the conservative nature of assumptions made about parts of the analysis of slope stability. The shear behaviour of soils typically shows peak strength at low to moderate strains and a lower residual strength. Using a value for the soil strength that is less than the peak will automatically be a safe assumption, and this in turn will underestimate the factor of safety.

Conversely, the actual analysis of slope stability involves engineering judgement to define a likely failure mechanism, which is usually done by delineating a potential failure surface. This is a mechanistic approach and will inevitably be an unsafe solution if an incorrect slip surface has been defined. Some spectacular failures on slopes that have been designed with care have failed for this reason, such as the example shown in Figure 2.6 and witnessed by the present author. A large, engineered cut slope failed, blocking one carriageway of the main highway west from São Paulo, Brazil in August 2002. A similar event occurred on Tsing Yi Island, Hong Kong in June 1982. The cause of this latter failure was an inappropriate use of a failure surface as a result of the designers not fully appreciating the underlying geological constraints.

Debate rages over the threshold factor of safety to be used, but less attention is often paid to a variable set of values that are determined by the consequence of failure. A defined factor of safety can always be achieved, but the cost of such action may not be justified if the consequence of a slope failure has limited impact on life or the local economy. An engineering approach may define a factor of safety for a particular slope, but for effective management, a variable set of factors is appropriate where the particular value is set according to the likely consequence. Thus, a higher factor of safety would be more relevant if the slope failure threatened a housing development in which people spend a significant part of their waking hours. On the other hand, it

Figure 2.6 Failure on an engineered slope at km 365 on the main highway west of São Paulo, Brazil.

would be difficult to justify such a high factor on most highways since the consequence of failure is less as it is associated not only with the failure itself, but with the probability that someone is passing at the precise time the landslide occurs. Intrinsic and extrinsic safety may thus be defined as follows:

Intrinsic safety: The factor of safety is determined for the slope in a traditional engineering approach without regard to the actual consequence. This will be a single value based on judgement and will always be greater than unity.

Extrinsic safety: The factor of safety is determined to a value that will vary according to the severity of the consequence as indicated above. This is the approach that was adopted by the Geotechnical Control Office in Hong Kong. The value for the factor of safety used in the extrinsic safety assessment may in some circumstances be only just above unity where the consequence of failure is very low, but will be significantly higher where the probability of loss of life, should failure occur, is high.

Adopting an extrinsic approach towards safety will automatically be a more cost-effective approach than an intrinsic definition of safety. However, if land use changes, then a false sense of security may occur, and care must be taken to re-appraise slopes in areas subject to such changes.

2. A GIS-based approach

The second approach usually adopts a GIS approach, which attempts to link general soil types, general slope angle and aspect, and general climatic conditions and the like with the historic incidence of landslides, as illustrated in Figure 2.7.

Landslide susceptibility maps often depict the likelihood of landslides in relative terms such as high, moderate or low, based on analyses or the weighing of factors contributing to slope instability. However, recent development of statistical analyses using GIS techniques have facilitated analyses of spatial data sets, resulting in graphical depictions of landslide potential in quantitative terms (Carrara and Guzzetti 1995; Guzzetti et al. 1999; Baban and Sant 2005). Such maps generally indicate where landslides are most likely to occur (Highland 1997; Guzzetti et al. 1999) but neither whether a specific slope will fail, nor when such

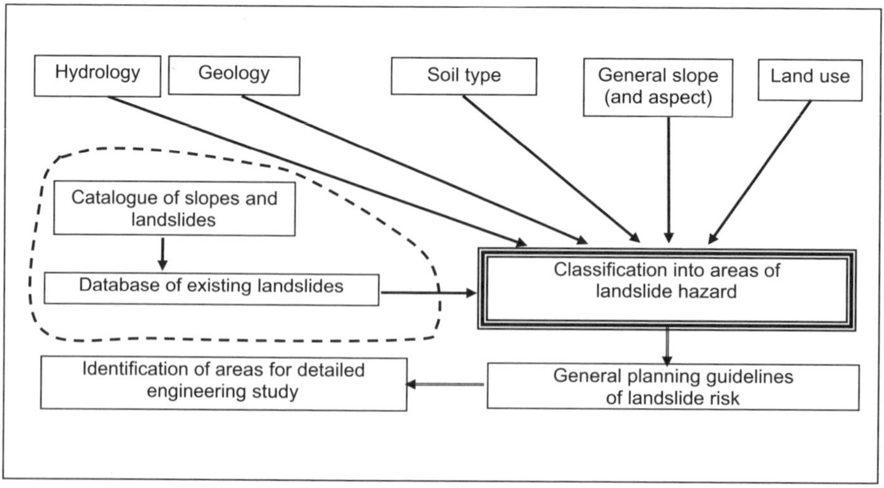

Figure 2.7 A GIS approach to analysis of slope stability.

failures will occur, as most hazard maps do not directly incorporate a time element.

Extensive databases of many key parameters, such as soil type, geology, land use, hydrology and the like are now available in some countries, making it possible to predict where the general failures might occur in the future. However, critical to this information is the need for a systematic database of slopes and previous landslides that have been accurately recorded, with respect to when they occurred and precisely where they occurred. Often this critical information is not available to the level needed in tropical countries. While this GIS approach is much less resource intensive than the engineering approach, it is also very much less accurate, both spatially and in time. Thus, it cannot specifically identify whether a particular slope might fail. Indeed it is only as good as the landslide database, and difficulty may exist in ensuring unbiased and complete reporting of all landslide incidents.

Figure 2.7 may be adequate initially, but it is usually deficient in several key areas. First, as noted previously, the likelihood of landslide occurrence is dependent on the type of slope, and information on this is rarely available, except in places such as Hong Kong. Second, there is no opportunity to include basic information on the mechanical properties of soils, which is fundamental to slope stability. It is true that soil type is a surrogate for this, but consideration should be given to include basic information on the mechanical properties as long as this can be achieved in a simple manner, such as that described later in this chapter.

3. A landslide warning system approach

The third approach attempts a correlation of historic landslide incidence with current and/or antecedent rainfall conditions as illustrated in Figure 2.8.

Information relating to the exact location of landslides, and the temporal and spatial incidence of rainfall, may be correlated for a given region to allow future predictions on the likelihood of significant landslide incidents. With this information, suitable warnings can be issued to the public and emergency teams can be mobilized effectively (Aliasgar and Baban 2006). However, while this may help to predict when landslides will occur, it cannot give information as to location. A

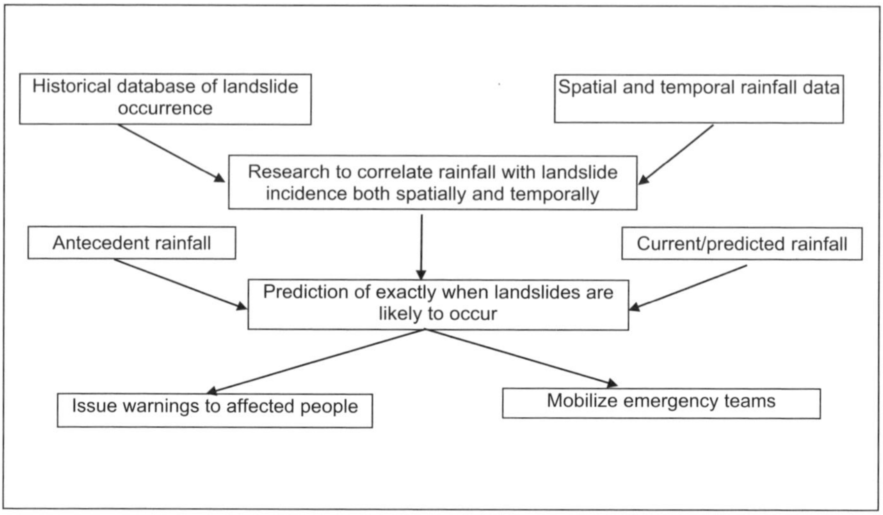

Figure 2.8 Steps in a landslide-warning system.

landslide warning system was first suggested by Lumb (1975), and developed further by the Geotechnical Control Office (GCO), and later renamed as the Geotechnical Engineering Office (GEO) (Premchitt 1984).

These early warning systems did have their faults, and the need to ensure that the correct infrastructure was in place to deal with such warning was highlighted by some spectacular errors of communication in the early days (particularly with respect to the rainfall event of 29 May–2 June 1982 [personal experience of the author]). Other researchers have also explored such predictive systems as to the incidence of rainfall-induced landslides in other parts of the world (Kay and Chen 1995; Fourie 1996; Toll 2001).

There are a few instances with respect to which the engineering approach will be important in tropical countries such as Trinidad and Tobago, particularly on key highways and in new developments on steep terrain. However, for effective use of resources, an adaptation of the second and third approaches is also likely to be of importance. To achieve this, it will be important to improve the field evidence of landslide occurrence and the GIS information available by including key engineering parameters such as shear strength estimated from simple tests, such as the Atterberg Limits and the Bulk Unit Weight, in a reference database. Furthermore, a clear appraisal of the types of slope, whether failed or not, is important, as is the mode of failure. Thus, in the central highlands of Trinidad, the principal landslide problem appears to be associated with fill slopes and retaining walls, while in the Northern Range, cut slope failures predominate. This chapter will explore how these developments can be combined to produce an effective, proactive landslide management scheme for the future and identify the critical further research that is needed. In particular, a collaborative approach involving research into all three approaches is important. Critical to this research are the resources needed to capture data for inclusion in a landslide database.

2.4 Proactive Management of Slopes

Landslides cause damage, injury, loss of life and economic loss, and a frequent response to such events is a reactive approach to deal with the consequences after the event. A proactive approach to slope management through risk assessment provides a rational basis on which to commit resources for landslide-preventative measures, and will, in the long term, provide a cheaper and safer solution to the hazard. Two major and serious landslides in Hong Kong in June 1972, in which over 140 people were killed (Schoustra 1972), provided a stimulus to move towards such a proactive approach in the management of the landslide hazards. In the late 1970s, a cataloguing of all slopes, whether failed or stable, was started, and this now includes over 50,000 slopes on a system that can be accessed over the Internet by the general public. This database now provides a rational basis for risk assessment for all such slopes and allows a rapid assessment to identify the most critical slopes through a ranking system. Such a system will be approximate and far from adequate to determine the true engineering stability of a slope, but it will, if designed correctly, provide a simple method to filter and identify the slopes most at risk. Thus a ranking system should incorporate key physical parameters such as slope height, slope angle and so on, and non-parametric ranking parameters such as condition, drainage, and consequence of failure. The ultimate aim is to obtain a single ranking factor, which can be achieved even with staff who possess limited experience.

If the slopes with the 100 highest scores are extracted from the database, then this group will almost certainly include the most critical slopes irrespective of any approximations that may be adopted in the aggregation of parameters in the ranking system. These 100 slopes can then be examined in more depth by experienced staff in order to identify the final group for which a full engineering analysis will be done. These selected slopes would then be the subjects of strengthening and other preventive measures in any one year, with other slopes selected in a similar way in following years. Such a system provides a rational basis for decision making for preventative measures, and ultimately, over a period of years, the most critical slopes should be strengthened against failure.

2.5 A Method for Slope and Landslide Recording in Trinidad

While hazard mapping, as outlined above, is possible, there can also be limitations, particularly in a country like Trinidad where the data on landslide occurrence is patchy at best. Developing GIS hazard maps using scant data may be of limited use. Furthermore, the focus of many studies has been to concentrate solely on known landslides with much less attention paid to those slopes, particularly those modified by man, which have remained stable over the recent past despite the presence of major rainstorm events. In any hazard mapping, information that a slope has not failed is of equal value to information about failed slopes. A cataloguing system similar to, or developed from, the one that was used in the Hong Kong System is thus important for research into the potential of landslide hazard and for the future management of consequence/mitigation of slope failures.

In many situations, the response of the authorities to major landslide incidents is reactive, and little consideration is given to the systematic recording and collation of such valuable information. Thus, on 9 December 2004, no fewer than 59 landslides on the North Coast Road were reported by the media (TV6 news broadcast, 9 December 2004).

However, no systematic recording of location or time was done by the authorities, and new and effective methods are needed to capture such information effectively without the need for excessive resources.

At present, there is no centralized system for recording landslide information in Trinidad and Tobago, and much of the information that does exist has been captured in a piecemeal approach. It is thus appropriate to consider the most effective approach for Trinidad, which can be drawn from looking at the success of different approaches in other countries. In addition, it is important to consider which, if any, of the three methods of landslide analysis is most appropriate for Trinidad at the present time and in the foreseeable future. In the past, there have been limits on the development of land above the 300 m contour, but recently there has been increased pressure for development above this level in the region surrounding Port of Spain and on either side of the Northern Range. Such development is associated with the building of appropriate infrastructure, such as roads, which themselves involve further human-influenced modifications to the slopes.

Where any such new development takes place, it makes sense to place adequate geotechnical control on all designs to ensure that, with regard to slope stability, they achieve an appropriate minimum extrinsic factor of safety. This control will normally require the testing of soils to ascertain key parameters such as the liquid and plastic limit, in situ moisture content, shear strength parameters and so on. Relevant slope stability analyses can then be done to check that an appropriate factor of safety is reached. This procedure can be costly to implement and is generally inappropriate for much existing development except perhaps in those areas deemed critical in a GIS analysis. However, when geotechnical data are obtained, they should be spatially recorded, to ensure that they can be geo-referenced for use in future GIS modelling.

The lack of a robust and systematic database of landslides in Trinidad is a major barrier to the effective research and development of successful management plans but, in the short-term, a system of management based initially on a GIS approach is likely to be the most beneficial. Some data does exist, but it is piecemeal, often housed in different departments, and it is by no means comprehensive or consistent. Indeed, after the establishment of the Geotechnical Control Office in Hong Kong in 1977, while the number of reported incidents at the time of major landslide events changed little, the number of reported minor landslide events increased by a factor of three to four times (personal experience by the author while working for GCO). This indicates a serious under-reporting prior to the establishment of a centralized coordinating body.

There are two important issues that must be addressed in the development of any landslide management system based around GIS: (1) the method to be adopted to catalogue and spatially locate slopes and landslides; and (2) how specific geotechnical information may be obtained effectively and how such information can be effectively integrated into the GIS analysis to produce landslide hazard maps.

Cataloguing and spatially locating slopes

The task to establish a coherent system for recording and cataloguing slopes and landslides can be daunting, and a simple and effective way of recording should be adopted that will be easy to develop and maintain once information obtained is incorporated into day-to-day management. Three different methods for recording information on the location and nature of slopes and landslides were investigated in Trinidad. Each of the methods has advantages, depending on the circumstances. However, all three methods must be capable of integration and expansion in a database, where additional information may be recorded. The three methods may be summarized as follows:

1. Recording using a simple and unique referencing system for each slope and landslide.
2. Recording slope and landslide features using GPS coordinates.
3. Recording slope and landslide features using road chainage markers, which are well established in most parts of Trinidad to the nearest 25 m.

The unique referencing method: The slope-referencing system adopted in Hong Kong is generally a robust method and is based initially on the map number at a scale of 1:20,000. In Trinidad, the relevant scale is 1:25,000. Within each map area, each slope feature is given a unique reference such that C001, F001, R001 and so on would be the first cut slope, fill slope and retaining wall catalogued in that area.

Opportunities for multiple features such as RF would indicate a retaining wall beneath a fill slope and so on. In Trinidad, a full reference would thus be of the following form: *43F-2b/FR0005*, where 43F represents the map 43F in the central highlands area and -2b the sub area of that map. FR005 indicates that the catalogued feature is the fifth one in that area and is a fill slope with retaining wall above. Such a code is concise, can be easily used for cross reference in a database and can conveniently be used in the field by reference to a hard copy of the map, on which the areal extent of the feature can be drawn while in the field.

Location and identification: Identifying landslide locations through using GPS coordinates related to the national grid coordinate system. These coordinates are readily converted into the relevant map area using suitable software and, thereafter, the coding system follows that using the map-based system. This approach is particularly useful for research using GIS methods, but it does require definitive GPS coordinate information, preferably by using differential GPS. Often, the emergency teams dealing with landslide events do not have access to such equipment.

Location and identification with reference to chainage points on roads: Most of the roads in Trinidad have painted markers at 25 m intervals, and this is a convenient way by which to record data. They are readily observable by anyone in the field and do not require access to a map or GPS facilities. Direct observation of position between the markers allows the positioning of landslide incidents to approximately 2 m to 5 m, which is suitable for all research work on landslide hazards. The main disadvantage of this approach is the need to have access to software to convert from chainage along a road to grid coordinates, although this facility is available in GIS packages. This method is particularly effective to capture information rapidly and is the method for reporting within the Highways Office. Two ways of capturing this data were explored. In the first, researchers walked the length of the road and recorded both slope type and landslide occurrence at the same time as shown in Figure 2.9. For a more rapid recording of landslide incidents following a rainstorm, time is of the essence, and it has thus been possible to drive along critical lengths of road, recording the location of each landslide shortly after an event. This provides a quick inventory,

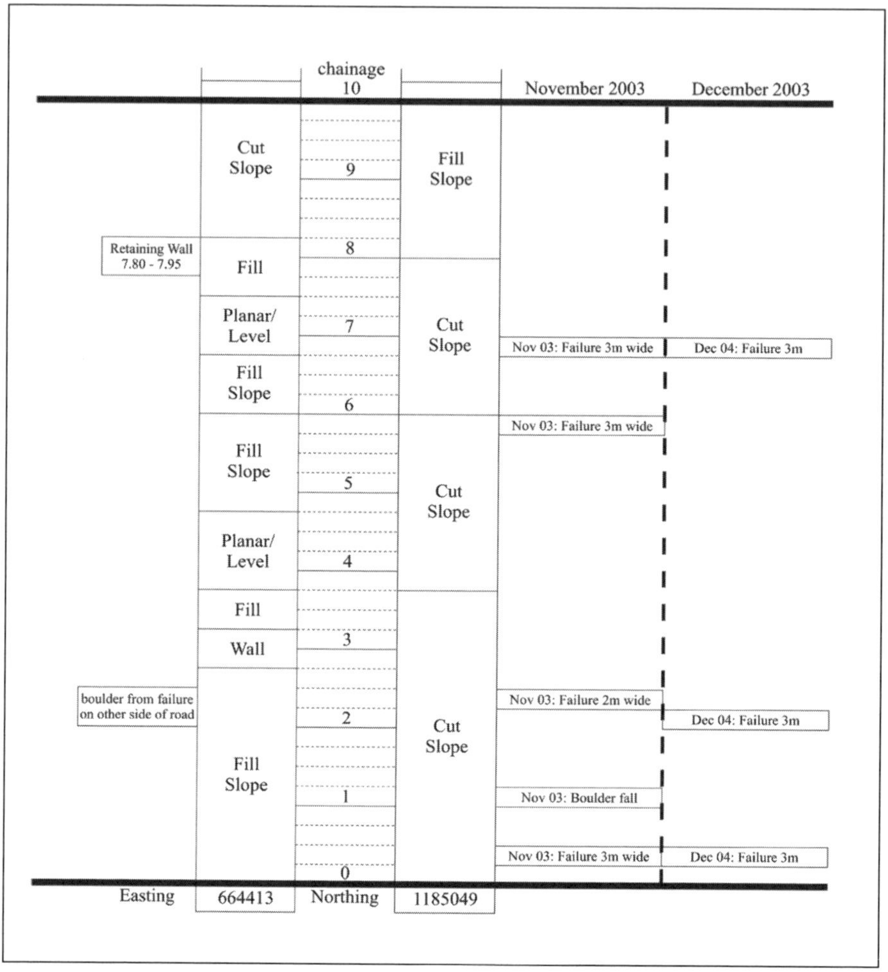

Figure 2.9 Example of booking using chainage as basis for spatial location. The chainage numbers refer to integral numbers of 100 m lengths from Maracas Junction.

allowing critical areas to be revisited later. Experience has shown that three researchers in a vehicle is optimum, driving along a stretch of road typically at 15 km to 20 km per hour. Other than the driver, one researcher can concentrate on identifying the chainage markers, another can identify the location and approximate size of any failure, and the other can record the data.

Where there are areas of particular significance, these areas can be completed on foot. An 18.5 km stretch of the North Coast Road between Maracas and Blanchisseuse was mapped in under two hours, on 20 November 2003, following a serious rainstorm. New landslides are clearly identifiable, but older landslides, which have occurred up to 10 years ago, are identifiable by obvious vegetation changes. On 11 December 2004, after a similar event, the full 30.8 km length of the North Coast Road, from Maracas Junction to Blanchisseuse, was mapped. On the first survey, both old and recent slides were recorded. On the second occasion, only evidence of landslides, which had occurred in the previous twelve months, was recorded.

Though the North Coast Road is notorious for landslides, of the total length of 30.8 km only 1,420 m of the north side of the road has been actually affected either prior to 2003, or in the two major events since. The corresponding figure for the south side was 1,670 m. Within the critical section from Maracas to Blanchisseuse, 75.2% of the length of road on the north side affected by landslides in December 2004 occurred either in areas of previous instability or within 10 m of the unstable areas. The figure for the south side was 72.9%. On the north side of the road, 83% of the landslides occurred over a short stretch of the road just 1.5 km long. Such information is important, since resources committed to landslide preventative measures at this specific location would be particularly effective in reducing the risk of landslide hazards in future.

Integrating geotechnical information into GIS analysis for landslide hazard analysis

Geotechnical data is rarely, if ever, incorporated into landslide hazard maps, and yet the shear strength of soils is one of the most important aspects in determining the stability of a slope. One approach would be to include the plasticity index, which is derived from the Atterberg Limits. Another approach could be to obtain an approximate estimate of the critical shear strength likely to occur at times of heavy rainfall from a knowledge of the liquid and plastic limits and the in situ porosity. Details of this approach are covered in Tovey (2006). The soil and geology digital maps may be used to identify where each unique combination of geology and soil type occurs, and thus locate where samples should be taken for geotechnical analysis.

The basic GIS flow diagram shown in Figure 2.7 may thus be modified to include this additional geotechnical information (Figure 2.10). If the hydrological information can also be extended to include not only spatial, but also temporal variations in rainfall as suggested by Campbell and Bernknopf (1997), then this could further enhance the predictive capability for landslide hazard mapping and provide a system which incorporates the better aspects of all of the three analysis methods discussed earlier. However, in addition to the capture of data on slope types, landslide occurrence and geotechnical parameters, a network of suitably located automatic rain gauges is needed. While the development of a full system may take quite sometime to achieve, the inclusion of geotechnical information is something that can and should improve the capability of existing systems at the present time.

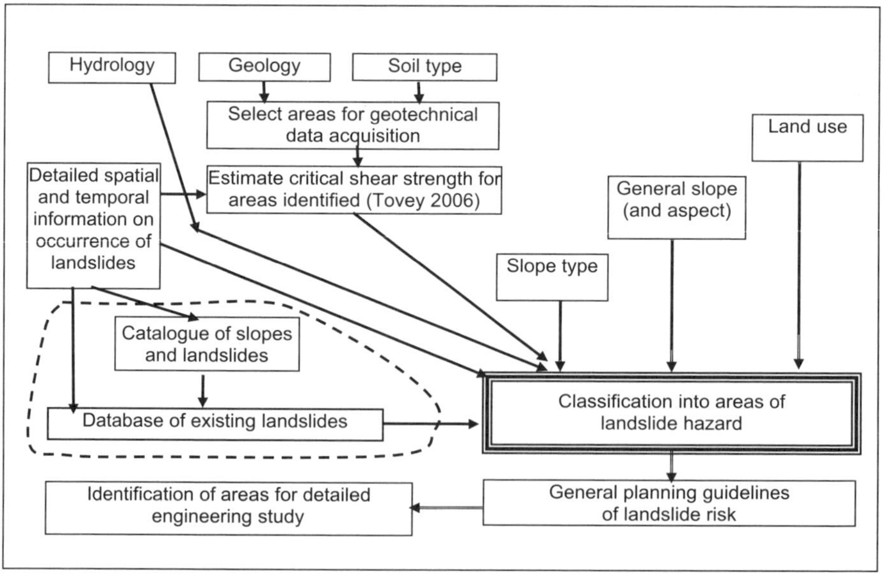

Figure 2.10 The GIS flow diagram of Figure 2.7 modified to include information from geotechnical measurements as suggested by Tovey (2006).

2.6 Conclusions

The management of landslide hazards in tropical countries may be effectively achieved using GIS methods. However, this requires that a robust and effective database and catalogue of landslides exists, in which the information has been recorded accurately with regard to both the time of occurrence and the location. In many tropical countries, even the basic information is lacking, and this chapter considers strategies to overcome these deficiencies in a country such as Trinidad. These include the following:

1. The development of an effective and efficient way in which to capture and record landslide and slope data. It is recognized that information on slopes which have not failed is of as much importance as information on landslides. Three different methods for this data capture and cataloguing are currently being explored:
   a. A system based on map area reference
   b. A system based on grid coordinates
   c. A system based on road chainage

   The last of these appears to be particularly effective both with respect to time and resources and is already providing some useful information. In particular, it is noted that on the North Coast Road, only a relatively small proportion is of serious concern, and approximately 75% of landslides appear to occur in areas that have been mobilized in the recent past.
2. It is important to try to bridge the differences in approach between the engineering, GIS and statistical methods and, where possible, to enhance the quantitative aspects of GIS methods since these will make more effective use of resources.
3. The importance of incorporating geotechnical information into GIS methods has been recognized, and an iterative procedure in which an initial GIS analysis identifies regions where simple geotechnical tests are done and is followed by the incorporation of such information in the final hazard mapping. Incorporating information based on aspects of the Atterberg Limits, together with predictions of the likely critical shear strength during periods of heavy rainfall, appears to be a promising way forward.

4. The present lack of a robust database incorporating precise information about the timing of occurrence of landslides means that statistically based analyses leading to a robust landslide warning system is difficult at the present time in Trinidad, although research should be conducted on the limited modelling facilities presently available to see if such information could be incorporated into an enhanced GIS analysis model.

Acknowledgements

This research was sponsored, in part, by a Higher Educational Link between the University of East Anglia, Norwich, United Kingdom, and the University of the West Indies, Trinidad.

References

Ahmad, R., and J.P. McCalpin. 1999. *Landslide susceptibility maps for the Kingston Metropolitan Area, Jamaica, with notes on their use*. UDS Publication no. 5. Kingston: Unit for Disaster Studies, Department of Geology, University of the West Indies.

Aliasgar, K., and S.M.J. Baban. 2006. Developing a geoinformatics based early warning system for landslides in Tobago. Paper presented at the Urban and Regional Information Systems Association Conference, The Bahamas. 30 October–2 November.

Baban, S.M.J., and K.J. Sant. 2005. Mapping landslide susceptibility for the Caribbean island of Tobago using GIS, multi-criteria evaluation techniques with a varied weighted approach. *Caribbean Journal of Earth Sciences* 38:11–20.

Campbell, R.H., and R. Bernknopf. 1997. Debris-flow hazard map units from gridded probabilities. In *Proceedings of the First International Conference on Debris-flow Hazards Mitigation: Mechanics, Prediction, and Assessment*, 165–75. San Francisco.

Carrara, A., and F. Guzzetti. 1995. *Geographical information systems in assessing natural hazards*. Dordrecht, Netherlands: Kluwer Academic Publisher.

Cruden, D.M., and D.J. Varnes. 1996. Landslide types and processes. In *Landslides: Investigation and mitigation*, ed. A.K. Turner and R.L. Schuster, 36–75. Transportation Research Board Special Report, no. 247. Washington, DC: National Academies Press.

Fourie, A.B. 1996. Predicting rainfall-induced slope instability. *Proceedings of the Institution of Civil Engineers: Geotechnical Engineering* 119, no. 4:211–18.

Gay D., 2004. Engineering approaches to landslide research in the Caribbean. Paper presented at the workshop Enduring Geohazards (Landslides and Floods) in the Caribbean Region, Learning Resource Centre, University of the West Indies, St Augustine, Trinidad. 8 December.

GEO. 1992. Reassessment of Po Shan landslide. Special Projects Division report SPR 16/92. Geotechnical Engineering Office, Hong Kong.

Guzzetti, F., A. Carrara and P. Reichenbach. 1999. Landslide hazard evaluation: A review of current techniques and their application in a multi-scale study, Central Italy. *Geomorphology* 31:181–216.

Highland, L.M. 1997. Landslide hazard and risk-current and future directions for the United States Geological Survey's landslide program. In *Landslide risk assessment,* ed. D.M. Cruden and R. Fell, 207–13. Rotterdam: Balkema.

Kay, J.N., and T. Chen. 1995. Rainfall-landslide relationship for Hong Kong. *Proceedings of the Institution of Civil Engineers: Geotechnical Engineering* 113, no. 2:117–18.

Lumb, P. 1975. Slope failure in Hong Kong. *Quarterly Journal of Engineering Geology* 8:31–65.

Premchitt, J. 1984. A review of landslip warning criteria. Special Project Division report SPR2/84. Geotechnical Control Office, Hong Kong.

Schoustra, J.J. 1972. *Po Shan Road landslip: Final report of the Commission of Inquiry into the rainstorm disasters*. Hong Kong: Government of Hong Kong.

Skempton, A.W., A.D. Leadbeater and R.J. Chandler. 1989. The Mam Tor landslide, North Derbyshire. *Philosophical Transactions of the Royal Society of London*, ser. A, no. 329:503–47.

Toll, D.G. 2001. Rainfall induced landslides in Singapore. *Proceedings of the Institution of Civil Engineers: Geotechnical Engineering* 149, no. 4:211–16.

Tovey, N.K. 2006. Incorporating geotechnical information into GIS landslide hazard mapping. (In preparation.)

CHAPTER 3

Planning for Hillside Terrains

DEBORAH THOMAS and SERWAN M.J. BABAN

Abstract

Today, there is a perceived scarcity of development land in Trinidad and Tobago in the face of an ever-increasing demand, especially for housing fuelled by rapid urbanization and escalating property prices. Consequently, there has been an unplanned rapid expansion of development into hillsides. This process is altering the natural slopes, resulting in the removal of protective natural vegetation, as well as the changing of the hydrological properties of catchments, leading to accelerated hillside erosion, landslides, floods and general environmental degradation.

The destructive impacts of accelerated expansion of development on hillsides are frequently underestimated, as statutory regulations agencies are poorly coordinated. Furthermore, the information bases used for development approvals are out of date, applications are dealt with in a piecemeal fashion and the current regulations are not enforced sufficiently.

Over the years, several initiatives have been taken to address the problems associated with development on hillside terrains in Trinidad and Tobago. However, most of these initiatives have been fraught with difficulties. There is still a need for a workable and enforceable appropriate policy framework to manage and guide development on hillsides in the interest of ensuring sustainable development.

In this context, the Ministry of Planning and Development appointed a Hillside Policy Technical Working Group in June 2004. This chapter reports on some of the outcomes from this initiative, which advanced a geoinformatics-based methodology for determining suitability for built development on hillsides, develop and implementing a simple scientific criterion strategy for managing hillside development in Trinidad and Tobago. Additionally, a case study is examined to highlight the applicability of the developed concept to Tobago. The results showed all potential areas suitable for hillside development in Tobago and also identified unsuitable areas and the reasons for disqualifying these areas.

3.1 Introduction

While public opinion may favour preservation of hillside areas in their natural conditions, it is necessary to balance the desire and need for preserving hillside areas with recognition of the need for development on a small island where land is a scarce and valuable resource. Hillsides, if managed properly, can play a critical role in realizing sustainable development and the well-being of society. Literature indicates that hillsides tend to serve a variety of functions (Chewing 1974; Nilsen et al. 1979; Erley and Kockelman 1981; Sidle et al. 1986; Thomas 2004; Baban and Sant 2007). These include the following:

1. *Residential and other built development:* Historically, people settled on hillsides in Trinidad and Tobago as far back as the post-emancipation era. Furthermore, the process of urbanization attracted people from rural to urban areas. They settled on hillsides around large cities to be close to where jobs and Crown land were available. Squatting, which is a widespread phenomena, is also a reflection of historical factors, continued urbanization, urban and rural poverty, homelessness, and landlessness. Today hillsides are prone real estate valued for their scenic news and symbols of prestige and wealth.
2. *Ecological:* Hillsides are habitat for wildlife and offer protection of bio-diversity. They are also important for forest conservation and play a critical role in habitat management and protection. The eco-

logical function of hillsides includes the protection of natural water ecosystems: streams, rivers, wetlands and coastal areas.

3. *Hydrological/watershed management:* Maintenance of vegetative cover and riparian corridors serve to sustain groundwater recharge and prevent degradation of water resources, including rivers, wells, springs and aquifers. Hillsides also function as part of the natural drainage system.
4. *Economic:* Hillsides provide valuable natural resources that are exploited for their commercial value. Quarrying and logging are common economic activities. Hillside land is also actively farmed, and guided tours provide a source of revenue from tourists. These activities provide sustainable livelihoods for residents of hillside communities.
5. *Aesthetics, recreation and culture:* Hillsides are attractive for their stunning views and valuable natural scenic qualities. They also provide numerous opportunities for passive recreation, biking, hiking, nature trails and guided tours. Consequently, hillside land has become not only a natural resource but also a valuable commodity that is desired and developed, sometimes unsustainably and often with serious consequences. Some of the issues associated with development on hillside terrains (Chewing 1974; Nilsen et al. 1979; Erley and Kockelman 1981; Sidle et al. 1986; Thomas 2004) include the following:

    a. Flooding and its associated impacts. These include death, damage to life and property, destruction of physical infrastructure, economic and financial losses, loss of agricultural crops, and other hardships. Flooding occurs in downstream locations due to:

        - increased housing and urban development that increase paved surfaces and result in increased runoff;
        - deforestation and removal of vegetative cover, again causing increased runoff and erosion; and
        - reduced capacity of drainage channels and water courses due to improper disposal of solid waste/garbage and increased runoff in drainage channels, which exceeds their design capacity.

b. Environmental degradation. This includes loss of vegetative cover, deforestation, loss of biodiversity, destruction of valuable habitats and sensitive environments to accommodate built development, and economic activity such as quarrying and improper agricultural practices, such as slash-and-burn, on slopes.
c. Landslides and slumping due to soil type, other soil characteristics, slope and geology.
d. Watershed degradation. Built development, quarrying, industrial activity and erosion in the upper catchment areas contribute to pollution of rivers and water sources. This results in deteriorating water quality, as evidenced by high levels of biological oxygen demand, bacterial content, turbidity due to sedimentation and the presence of chemical pollutants in rivers, and has serious implications for public health and ecosystem integrity.
e. Visual impacts. The scarring of hillsides caused by insensitive, careless and often unauthorized development negatively affects the visual and aesthetic appeal of our hillsides.
f. High cost of infrastructure provision to higher elevations and steep sites and maintenance of same.
g. Fire hazards during the dry season.

It is also important to realize if development is improperly planned, the very amenities that people seek can be destroyed. Therefore, the intervention in the land use and development process to achieve stated social, economic and environmental goals via effective planning – spatial or land-use planning – is critical.

By definition, stable steep slopes are in a state of equilibrium. When this equilibrium is disturbed due to natural or anthropogenic influences, including development in hillside areas, the likely consequences are often the removal of vegetative cover, which may contribute to erosion, slope failure, accelerated surface run-off and perennial flooding.

Hillside management programmes seek to define those areas that, because of their physical, environmental and functional significance, require varying degrees of protection and provide for areas where development and varying degrees of landform modification may occur.

Therefore, slope stability is a core issue in hillside development, and it is generally affected by three interrelated factors: water, geologic structure and lithology, and the area's gradient (Sidle et al. 1986; Baban and Sant 2007). Several approaches have been developed for estimating slope stability and vulnerability to landslides in a particular geographical area based on the existence of favourable factors, such as topography, geology, land use/cover and past history/inventory of landslides (OAS 1991; Baban and Sant 2004). These factors can be mapped and examined, and the conditions present in an area can then be factored together to represent the degree of potential vulnerability present (DeGraff and Rhomesburgh 1980).

The literature indicates clearly that in establishing hillside management programmes the following determinants need to be considered: slope, accessibility, cost of public services, natural and aesthetic resources, waster disposal and geohazards (Chewing 1974; Nilsen et al. 1979; DeGraff et al. 1989; Erley and Kockelman 1981; Sidle et al. 1985; Moser 1991; Marsh 1991; LSA Associates 2002).

Identifying land suitability for specific applications, which is a critical necessity for rapidly developing small island states, is becoming a science of its own as, among other things, decision makers now have to understand the geology, hydrogeology and ecology as well as cultural attributes of sites. The shortage in reliable and accurate data sets is also a critical challenge in developing nations (Baban 2004). However, these problems can be managed in the Caribbean region by developing practical scientific criteria based on proven experience worldwide and by using geoinformatics, which comprises the necessary tools – such as geographic information systems (GIS), remote sensing and global positioning systems (GPS) – to collect, manipulate and analyse data, thereby overcoming the information poverty issues (Baban et al. 2004).

3.2 Managing Hillside Development in Trinidad and Tobago, Challenges and Opportunities

In Trinidad and Tobago today, there is a perceived scarcity of development land in the face of an ever-increasing demand, especially for housing fuelled by rapid urbanization and escalating property prices. While there are obvious attractions to hillside locations, the increasing demand for hillside land for housing around Port of Spain may be attributed to continued unsustainable patterns of development which concentrate employment opportunities in and around the capital city, while the workforce commutes from dormitory settlements in the east-west corridor and central Trinidad, and even from towns and villages further away. This demand is therefore unlikely to be a reflection of any intrinsic value of hillside locations and may be more a desire to spend less time in traffic during the daily commute to and from work in Port of Spain (Thomas 2004).

The apparent proliferation of development on hillside terrains in Trinidad and Tobago has been highlighted, due to death and damage in recent years caused during natural hazard events locally and elsewhere in the region. In particular, landslides in Tobago have been reportedly responsible for two deaths on the island, injury to residents and significant damage to homes, property and the environment (Thomas 2004; Baban and Sant 2005). In recent times, many valleys in the Northern Range of Trinidad experienced floods for the first time in living memory, leading to damage to property and infrastructure, and disruptions to transport and the social fabric.

The negative impacts of hillside developments tend to be underestimated, as development control agencies that deal with applications for statutory approvals are poorly coordinated, and applications are dealt with in a piecemeal fashion. Additionally, the information on which decisions are made regarding development approvals is out of date and lacks detail relating to existing conditions with the situation on the ground, and susceptibility information is generally lacking. Finally, the lack of enforcement of existing environmental protection laws is seen as a significant contributor to the uncontrolled and accelerated expansion of development within hillside areas. The erection of one dwelling in an area often leads to others (Thomas 2004). In the context of the Government of Trinidad and Tobago's Vision 2020 development strategy and the proposed Local Government Reform, present developmental patterns and trends neither appear to support the concept of more balanced regional development, nor the creation of sustainable communities.

The legal framework to regulate and manage development is the

Town and Country Planning Act Chapter 35:01 and the Environmental Management Act 1995, amended in 2000. The former requires obtaining the permission of the minister responsible for town and country planning prior to carrying out development as defined in the act, while the latter requires developers to obtain Certificates of Environmental Clearance (CEC) under the Designated Activities Order and CEC Rules from July 2001 for specified categories of development. The persistent failure of developers to adhere to and comply with laws denies regulatory agencies the opportunity to evaluate developmental proposals and recommend appropriate mitigation measures.

Meanwhile, the unplanned rapid expansion of development into hillsides is ongoing and has the effect of altering the natural slopes as well as the hydrological properties of catchments. Furthermore, agricultural and residential squatter settlements on hillsides are resulting in the removal of protective natural vegetation and loss of wildlife habitat. The common practice of slash-and-burn agriculture on small farms is also contributing to accelerated hillside erosion and general environmental degradation, as it results in land clearance and the eventual abandonment of farms. Consequently, this barren land is more susceptible to landslides and increased runoff, which may in turn lead to flooding in the lower elevations. Therefore, a workable and appropriate policy framework is necessary to manage and guide development on hillsides in the interest of ensuring sustainable development. Due to the aforementioned concerns, several planning and policy initiatives have been taken over the years to address the problems associated with development on hillside terrains. In general, the initiatives seek to protect lives and property; optimize the use of natural resources; preserve scenic views, aesthetic value and the quality of the natural environment; and conserve biodiversity.

This can be achieved by attempting to determine what, if any, is the most appropriate use for hillside land: the appropriate conditions and design and development standards to be applied in order to meet the needs for safe and affordable housing and other forms of development, while respecting and protecting human lives, property and the natural environment, minimizing erosion, landslides, flooding and other negative impacts, as well as preserving the aesthetic quality of our hillsides.

The literature indicates (Chewing 1974; Nilsen et al. 1979; Erley and Kockelman 1981; Sidle et al. 1986; Thomas 2004; Baban and Sant 2007) that this may be achieved by any or a combination of the following techniques and strategies:

1. Defining slopes and elevations above which development may be prohibited or restricted.
2. Determining the carrying capacity of hillsides to accommodate additional development, especially in relation to available physical infrastructure.
3. Regulating density – for instance, by reducing densities on steeper slopes, establishing minimum plot sizes, specifying a percentage of the development site to be retained under natural vegetation or restricting the maximum number of buildings permitted on slopes.
4. Regulating development on the basis of soil type and other soil characteristics; land capability for agriculture and geology.
5. Establishing guiding principles and performance standards for developers in terms of design, setbacks, shared access, building height, massing, clustering, reduced road widths and so on. This may be applied with some degree of flexibility.
6. Environmental management principles to regulate cutting and grading of slopes to avoid scarring of the landscape, erosion, drainage problems and unnecessary loss of vegetation.
7. Identifying appropriate mitigation measures to reduce the long-term vulnerability of human life and property to the negative impacts of hillside development. The CEC and environmental impact assessment processes of the Environmental Management Authority play a significant role in this context.
8. Integrated approaches at the institutional level to facilitate coordination of the multiplicity of public, private and civil society organizations.
9. Public education and awareness to sensitize the general population and developers (both the public and private sector) to the issues and impacts of hillside development.

It is recognized that planning policies and standards may reduce the viability of hillside development and restrict the use of private lands. In some situations, flexibility in the application of policies and standards is necessary to reduce hardship on landowners. However, at times acquisition and compensation may be the most desirable option in the interest of achieving sustainable development. According to Thomas (2004), the following policy measures were initiated to address the problem in Trinidad and Tobago:

1. Cabinet Minute no. 692 of 1976 restricted development in Trinidad and Tobago above the 30-foot contour (largely because of the inability of the Water and Sewerage Authority [WASA], at the time, to pump water to higher elevations) and on slopes steeper than 1:6. This policy was considered limited, restrictive and inappropriate.
2. Cabinet Minute no. 1995 of 1979 authorized appointment of a committee to prepare a hillside development policy and translate it into a development plan. The Northern Range Hillside Development Policy was produced in 1988. This policy has formed the basis for guiding development on hillside terrains and has sought to preserve soil stability, maintain ecological and hydrological balance, and facilitate sustainable land-use choices. However, it has been criticized for inconsistency in its application, lack of compliance and enforcement, generality of application, and restrictive focus on slope stability and gradient to the apparent exclusion of other relevant factors.
3. Preparation of spatial development plans for Trinidad and Tobago as whole and critical valleys of the Northern Range and other parts of Trinidad and Tobago was sought to guide and manage development on the basis of suitability, carrying capacity and sound environmental principles.
4. Other policies have sought to protect forest reserves for soil conservation, habitat preservation and watershed protection.

Unfortunately, unplanned development is ongoing in the hillsides. Evidently, ineffective regulations and policies, as well as attitudes regarding the environment and development have created difficulties for realizing sustainable development. An added difficulty in the Caribbean region is the information poverty phenomena which constrains informed decision making (Baban et al. 2004).

In this context, several initiatives are being developed. The Minister of Planning and Development appointed a Hillside Policy Technical Working Group in June 2004 to

1. articulate the issues surrounding development of hillsides in Trinidad and Tobago;
2. review existing policy and standards on hillside development to determine the necessary revisions required in the light of increased demand and new technologies in the development of hillsides;
3. facilitate collaboration with the Town and Country Planning Division and other key stakeholders to formulate appropriate developmental standards and to design guidelines to regulate and manage development on hillsides in the interest of sustainable development; and
4. seek consensus, endorsement and compliance with respect to the agreed policy of all stakeholders.

As a result, the group advanced a methodology for determining suitability for built development on hillsides, and developed and implemented a scientifically based, geoinformatics-driven criteria for managing hillside development in Trinidad and Tobago, while enduring information poverty. Additionally, a case study was presented to highlight the applicability of the concept to Tobago (Baban et al. 2006). The following section will report on some of the outcomes from this project.

3.3 Developing a Composite Criterion for Managing Hillside Development in Trinidad and Tobago

3.3.1 Criteria Development

The siting criteria used for managing hillside development by various organizations worldwide contain very similar elements. These are focused on the physical and environmental characteristics of investigated areas, such as land capability, slope, hazard and risk or spatial attributes, such as proximity to the road and settlements. Naturally, each organization has adjusted its respective criteria to reflect its individual needs. Based on the literature and expertise of the technical

Table 3.1 Hillside Suitability Criteria

Analysis Level	Criteria	Constraints Factor	Consideration
Level I Catchment-level analysis	(1) Avoid critical catchments	Hydrological	Environment
Level II Site-specific analysis, stage 1, stage 2	Avoid protected areas	Ecological	Environment
	(1) Avoid taking land capability class <= IV	Agriculture land capability	Natural resources
	(2) Be located within 1 km to the present settlement	Social and access	Planning
	(3) Be located within 1 km to the major road	Access	Planning
	(4) Avoid flood and landslide hazard area	Risk	Planning
	(5) Avoid terrain slope < 1:3	Physical	Planning

working group (TWG), a geoinformatics-based criteria for hillside development in Trinidad and Tobago was established (Table 3.1). The proposed criteria (Table 3.1) show the analysis level, constraint criteria, constraints factor and the consideration (highlighted as being planning, environmental or resource based). Constraints criteria have been organized as a set of Boolean rules, that is the site must be located so that all of the conditions are satisfied.

3.3.2 Methodology

A methodology for developing a geoinformatics-based system to implement the established suitability criteria for built development was developed. The system is designed to function at two levels: the hydrological catchment level and the specific potential site level (Figure 3.1).

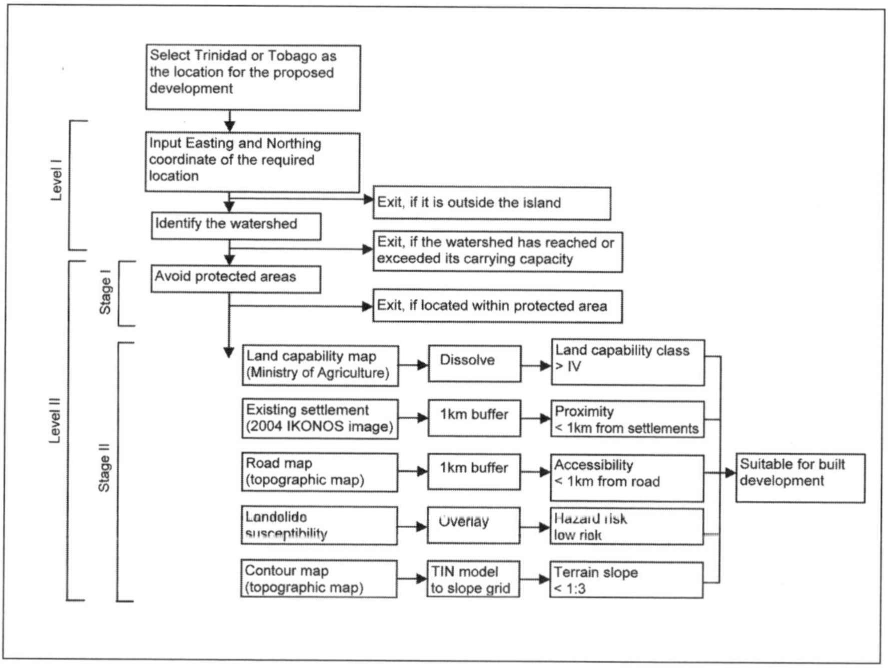

Figure 3.1 A flow diagram of the geoinformatics-based system for managing hillside development (after Baban et. al 2006).

3.3.3 Implementing the Criteria: A Case Study of Tobago

The criteria established by the TWG were applied to Tobago to develop and evaluate a geoinformatics-based planning support system for hillside development on the island. A questionnaire was developed to collect all relevant existing data coverages, which directly or indirectly provided information about the specific criteria for built development, such as coverages on the following themes: topography, critical slopes,

Table 3.2 Digital Data Sets Developed for Tobago

Data Layer	Data Form	Scale	Source Agency
Coastline	Paper map	1:10,000	Directorate of Overseas Surveys
Watershed	Digital data	–	Water and Sewage Authority
Land use/cover	Digital data	–	CLEAR, UWI
Protected area	Paper map	1:100,000	Ministry of Agriculture
Land capability	Paper map	1:25,000	Ministry of Agriculture and UWI
Settlement	Digital data	–	CLEAR, UWI
Road	Paper map	1:10,000	Directorate of Overseas Surveys
Landslide	Digital data	–	CLEAR, UWI
Contour	Paper map	1:10,000	Directorate of Overseas Surveys

land use/land cover, watersheds, agricultural land capability, settlements/built development, hazard/risk vulnerability, road networks, and environmental and ecologically-sensitive areas. The data was compiled, evaluated and developed into a geo-database. Based on the information obtained, the data sets were acquired from the agencies and customized according to the requirements of the project. These are listed in Table 3.2.

A geoinformatics-based system using established criteria to identify hillside land suitable for built development was developed (Figure 3.1). Level I analysis was applied to all the watersheds in Tobago (Figure 3.2). To fulfil Level II analysis, the following procedure was used: the digital land capability map was dissolved into two classes. The first class was less than equal to capability class IV, which was suitable for agriculture, and therefore it is not suitable for built development (Figure 3.3). The second class was greater than capability class IV, which is suitable for built development. For each class, 1 and 0 values

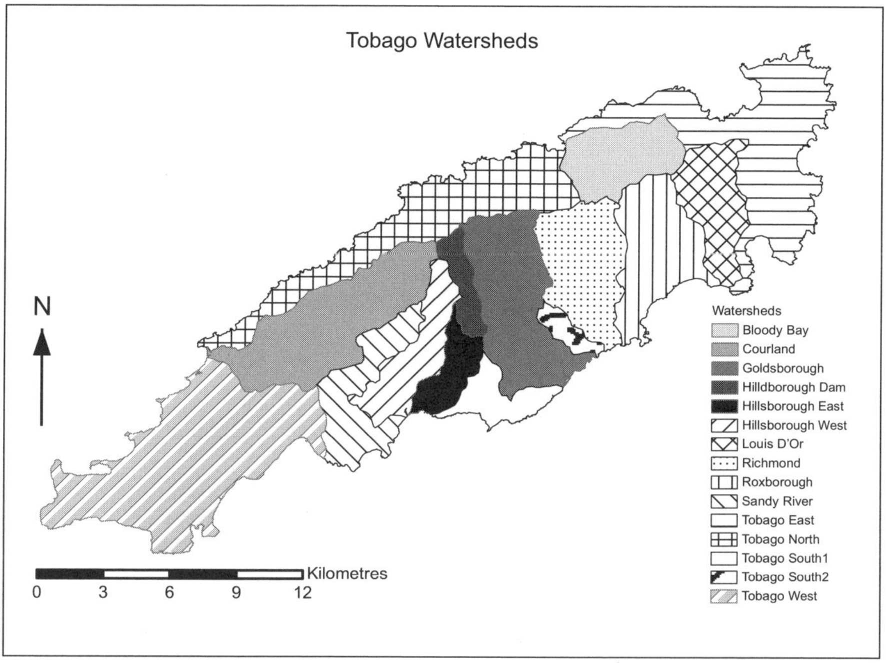

Figure 3.2 Hydrological catchments in Tobago.

were assigned for suitable and unsuitable respectively. A distance of 1 km area from the existing settlement extracted from the satellite image was buffered, considered for built development and value 2 was assigned. The rest of the areas that were not suitable were given 0.

The road map was buffered to 1 km distance from the existing road. Based on the assumed 1 km distance, the area within 1 km is suitable and the value 4 was assigned, and the area beyond 1 km is unsuitable and 0 was assigned. Areas vulnerable to landslides are unsuitable; therefore, the value 0 was assigned to these areas while all other areas were given the value 8. The triangulated irregular network (TIN) developed from the topographic contour map was used to derive slope values. The outcome slope was dissolved into two classes to show slopes equal to and exceeding 1:3, and, of the rest, the value 0 was allocated to the first class and 16 to the second.

All these layers were geo-referenced, overlaid and added according to the placement of those values. Finally, from the added values, the

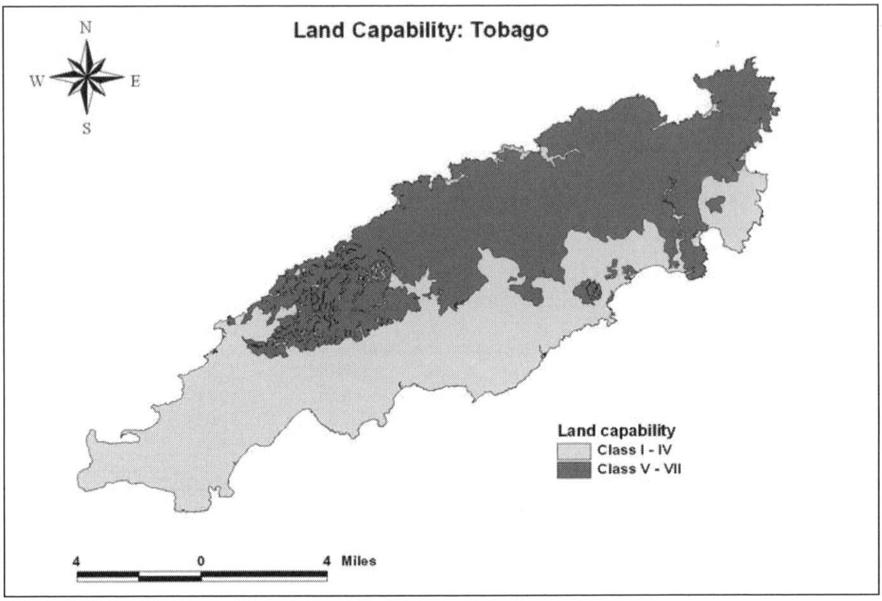

Figure 3.3 Agricultural land capability.

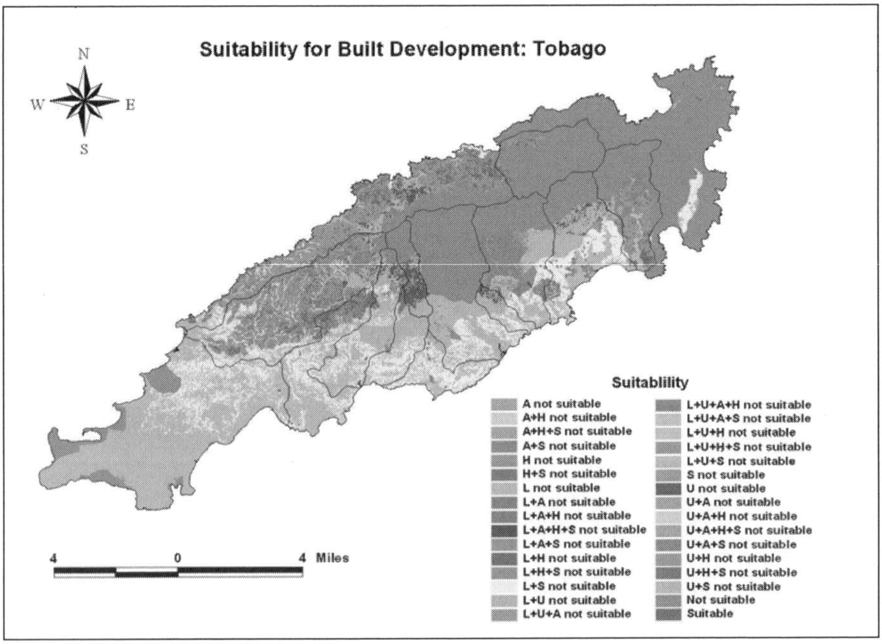

Figure 3.4 Land suitability for built development – Tobago.

class types were determined (Figure 3.4). The final outcome shows areas suitable for hillside development in Tobago based on fulfilling all the criteria used. Furthermore, areas that are not suitable and the reasons for disqualifying these areas are also identified (Table 3.3).

Table 3.3 Land Suitability Criteria

Total Value	Class	Description
0	Not suitable	All criteria not satisfied
1	U+A+H+S not suitable	Unsuitable due to proximity to the settlement (U), accessibility (A), hazard (H) and slope (S)
2	L+A+H+S not suitable	Unsuitable due to land (L), accessibility (A), hazard (H) and slope (S)
3	A+H+S not suitable	Unsuitable due to accessibility (A), hazard (H) and slope (S)
4	L+U+H+S not suitable	Unsuitable due to land (L), proximity to the settlement (U), hazard (H) and slope (S)
5	U+H+S not suitable	Unsuitable due to proximity to the settlement (U), hazard (H) and slope (S)
6	L+H+S not suitable	Unsuitable due to land (L), hazard (H) and slope (S)
7	H+S not suitable	Unsuitable due to hazard (H) and slope (S)
8	L+U+A+S not suitable	Unsuitable due to land (L), proximity to the settlement (U), accessibility (A) and slope (S)
9	U+A+S not suitable	Unsuitable due to proximity to the settlement (U), accessibility (A) and slope (S)
10	L+A+S not suitable	Unsuitable due to land (L), accessibility (A) and slope (S)
11	A+S not suitable	Unsuitable due to accessibility (A) and slope (S)

Table 3.3 continues

Table 3.3 Land Suitability Criteria *(cont'd)*

Total Value	Class	Description
12	L+U+S not suitable	Unsuitable due to land (L), proximity to the settlement (U) and slope (S)
13	U+S not suitable	Unsuitable due to proximity to the settlement (U) and slope (S)
14	L+S not suitable	Unsuitable due to land (L) and slope (S)
15	S not suitable	Unsuitable due to slope (S)
16	L+U+A+H not suitable	Unsuitable due to land (L), proximity to the settlement (U), accessibility (A) and hazard (H)
17	U+A+H not suitable	Unsuitable due to proximity to the settlement (U), accessibility (A) and hazard (H)
18	L+A+H not suitable	Unsuitable due to land (L), accessibility (A) and hazard (H)
19	A+H not suitable	Unsuitable due to accessibility (A) and hazard (H)
20	L+U+H not suitable	Unsuitable due to land (L), proximity to the settlement (U) and hazard (H)
21	U+H not suitable	Unsuitable due to proximity to the settlement (U) and hazard (H)
22	L+H not suitable	Unsuitable due to land (L) and hazard (H)
23	H not suitable	Unsuitable due to hazard (H)
24	L+U+A not suitable	Unsuitable due to land (L), proximity to the settlement (U) and accessibility (A)
25	U+A not suitable	Unsuitable due to proximity to the settlement (U) and accessibility (A)
26	L+A not suitable	Unsuitable due to land (L) and accessibility (A)

Table 3.3 continues

Table 3.3 Land Suitability Criteria *(cont'd)*

Total Value	Class	Description
27	A not suitable	Unsuitable due to accessibility (A)
28	L+U not suitable	Unsuitable due to land (L) and proximity to the settlement (U)
29	U not suitable	Unsuitable due to proximity to the settlement (U)
30	L not suitable	Unsuitable due to land (L)
31	Suitable	All criteria are satisfied

In this case, the criteria were applied to all the hydrological catchments in Tobago, assuming that none of the catchments has reached or exceeded its carrying capacity. This decision was made with support from the TWG due to lack of information regarding the actual carrying capacity status of hydrological catchments in Tobago. Table 3.4 illustrates all 15 watersheds of Tobago, the size of the watersheds, suitable areas for built development within the watershed and the percentage of suitable area per total area. The total area of Tobago is approximately 74,054 acres, of which 3,780 acres (5%) are suitable for built development. The Tobago North and Courland watersheds have a high potential for built development. When considering percentage, these two watersheds show more than 17%, and they may be critical for built development, as present land cover will be disturbed by built development. Furthermore, Sandy River watershed shows some suitable areas for built development. However, it should be noted these areas are located in the upper catchments (Figure 3.4).

When taking into account the size of the watersheds and their suitability for development, there is a clear need to improve the unsuitable areas for built development, where appropriate, in order to optimize land use. Statistical analysis for all the watersheds in Tobago shows the watersheds and their total area, protected areas, and areas not suitable for built development based on the TWG criteria. Over 23,700 acres of

Table 3.4 Watershed Suitability for Development in Tobago and Their Associated Area Statistics

Watershed	Total Area (Acres)	Suitable Area (Acres)	Percentage
Tobago East	9,627	67	1
Bloody Bay	3,611	110	3
Louis D'Or	3,513	256	7
Tobago North	7,745	1,303	17
Roxborough	4,970	160	3
Richmond	4,932	50	1
Goldsborough	6,495	–	–
Hillsborough Dam	1,247	13	1
Courland	7,556	1,318	17
Hillsborough West	3,188	58	2
Tobago South	2 739	–	–
Sandy River	3,787	394	10
Hillsborough East	1,758	50	3
Tobago South	11,828	–	–
Tobago West	13,058	–	–
Total	**74,054**	**3,780**	**5**

natural reserve are identified as protected areas, and more than half of Tobago East, Bloody Bay and Goldsborough watersheds are protected areas. Around 44% of Tobago's land is best suitable for agriculture and so fail to satisfy TCW criteria. Land best suited for agriculture is also well suited for built development since the criterion for agricultural land capability included gentle topography.

Therefore, there is an inherent conflict that must be managed. This requires articulation of a national policy on the importance of agriculture to the economy of the country and its contribution to gross domestic product. It is evident that slopes are one of the main criteria in

Table 3.5 Tobago Watersheds and Reasons for Unsuitability

Watershed	Total Area (Acres)	Protected areas	Agriculture Land Not Suitable Cap. <=IV	Far from Settlement Not Suitable > 1km	Accessibility Not Suitable > 1km	Hazard Not Suitable	Steep Slope Land Not Suitable > 1:3
Tobago East	9,627	8,917	433	50	55	11	510
Bloody Bay	3,611	3,297	–	–	–	1	204
Louis D'Or	3,513	1,687	273	482	590	61	1,235
Tobago North	7,745	1,259	819	1,126	309	64	4,233
Roxborough	4,970	1,161	2,493	1,153	13	79	2,390
Richmond	4,932	1,556	1,980	2,151	1,881	24	2,218
Goldsborough	6,495	3,648	2,764	1,067	278	8	1,498
Hillsborough Dam	1,247	502	180	728	–	6	187
Courland	7,556	–	2,278	859	260	54	4,705
Hillsborough West	3,188	–	2,644	920	34	5	1,250
Tobago South 2	739	2	592	362	–	12	344
Sandy River	3,787	–	3,091	31	–	13	1,384
Hillsborough East	1,758	–	1,680	498	9	5	844
Tobago South 1	1,828	–	1,827	101	–	46	982
Tobago West	13,058	1,495	11,556	–	–	4	1,976
Tobago	**74,054**	**23,758**	**32,609**	**9,478**	**3,429**	**395**	**23,959**

Tobago as it makes 32% (23,959 acres) of the area not suitable for built development. The issue of slope must be considered in the context of soil type and vulnerability to landslides, which restrict suitability for development, and the cost of providing infrastructure on steep slopes.

A significant proportion of land in Tobago (9,478 acres are far from the present settlements, and 3,429 acres are not accessible), could be utilized for built development subject to creating new infrastructure in those areas (Table 3.5). Richmond watersheds show high potential for built development if the new infrastructure is installed (Table 3.5). Tobago North and Courland watersheds show high potential for built development, and Richmond shows the potential if infrastructure is provided within the watershed (Table 3.4 and Table 3.5).

3.4 Conclusions and Recommendations

The development of scientific criteria to objectively manage development in the Caribbean region and in Trinidad and Tobago is critical to the achievement of sustainable development and "developed country" status.

Major land-use changes have been occurring in Trinidad and Tobago over the past many years, fuelled by urbanization, development, economic growth, globalization, trade liberalization and other factors. While natural hazard events have always been present in the region, disasters occur when a natural hazard event intersects with the human-influenced environment. Consequently, there is a need to connect the dots and make the connection between human activities such as hillside development and the consequences of such actions.

The Town and Country Planning Division in Trinidad and Tobago has embarked on the preparation of a Revised National Physical Development Plan, incorporating hazard identification and mitigation measures and including land-use proposals based on site suitability, carrying capacity and other sustainability measures. A new spatial strategy for Trinidad and Tobago will also be based on similar principles. These circumstances have fuelled the need for the Government of Trinidad and Tobago to develop a comprehensive, transparent and impartial National Hillside Development Policy. The Town and Country

Planning Division established a TWG during 2004 to facilitate the development of the National Planning Policy for Hillside Development.

The TWG developed a geoinformatics-based approach for managing hillside development in Trinidad and Tobago. This approach promotes the use of geoinformatics to manage the information poverty challenge through the development of the missing but essential data sets, as well as carrying out the necessary analysis. The applicability of the concept was examined through a case study in Tobago.

The proposed approach is flexible, which makes it useful as a planning tool as it provides the planners with the freedom to employ their individual, local, national and regional expertise in the decision making process. Additional relevant transient layers of information, such as public satisfaction, could be easily integrated into the approach and, consequently, be taken into consideration when required. Furthermore, the recommended approach is also scientifically justifiable, open to scrutiny and able to lend itself to public acceptance in the future. This approach also indicates reasons for disqualifying developmental proposals. This is critical as some of these reasons can be overcome by engineering solutions. This approach could be taken a step further by using GIS to assist in managing development by locating an optimum site among several of the "most suitable" sites from the constraint map and assessing their suitability on an individual basis. Moreover, it is also possible to determine the optimal size and number of development units in each geographical area, based on potential natural growth as well as on demographic shifts in population due to employment and other related factors.

Acknowledgements

The authors are grateful to the Town and Country Planning Division, Ministry of Planning and Development, Trinidad and Tobago, for sponsoring this research and to members of the Technical Working Group for their valuable contribution in developing and implementing Composite Criteria for Managing Hillside Development in Trinidad and Tobago.

References

Baban, S.M.J. 2004. Attaining a balance between environmental protection and sustainable development in the Caribbean region using geoinformatics. *West Indian Journal of Engineering* 26, no. 2:22–34.

Baban, S.M.J., F. Canisius, K. Sant and A. Chinchamme. 2006. Technical inputs to the Hillside Development Policy Technical Working Group. Ministry of Planning and Development, Government of the Republic of Trinidad and Tobago.

Baban S.M.J., B. Ramlal and R. Al-Tahir. 2004. Issues in information poverty and decision-making in the Caribbean region: A way forward. *West Indian Journal of Engineering* 27, no. 1:28–37.

Baban, S.M.J., and K.J. Sant. 2004. Mapping landslide susceptibility on a small mountainous tropical island using GIS. *Asian Journal of Geoinformatics* 5, no. 1:33–42.

———. 2005. Mapping landslide susceptibility for the Caribbean island of Tobago using GIS, multi-criteria evaluation techniques with a varied weighted approach. *Caribbean Journal of Earth Sciences* 38:11–20.

———. 2007. Identifying critical slopes for landslide management in mountainous tropical environments using geoinformatics. *West Indian Journal of Engineering*. In press.

Chewing, J.A. 1974. *Hillside studies and legislation across the United States*. Cincinnati: The Cincinnati Institute.

DeGraff, J.V, R. Bryce, R.W. Jibson, S. Mora and C.T. Rogers. 1989. Landslides: Their extent and significance in the Caribbean. In *Landslides: Extent and economic significance*, ed. E.E. Brabb and B.L. Harrod, 51–80. Rotterdam: Balkema.

DeGraff, J.V., and H.C. Rhomesburgh. 1980. Regional landslide susceptibility assessment for wild land management: A matrix approach. In *Thresholds in geomorphology, Binghampton Symposium in Geomorphology*, ed. C.R. Coats and J. Vitek, 401–15. London: Allen and Unwin.

Erley, D. and W.J. Kockelman. 1981. *Reducing landslide hazards: A guide for planners*. Planning Advisory Service Report no. 359, Chicago: American Planning Association.

LSA Associates. 2002. Overview of hillside development terms and concepts.

Marsh W.M. 1991. *Landscape planning: Environmental applications*. 2nd ed. New York: John Wiley and Sons.

Moser W.A. 1991. *Design for successful hillside development. ASCE Journal of Urban Planning and Development* 117, no. 3:85–94.

Nilsen, T.H., R.H. Wright, T.C. Vlasic and W.E. Spangle. 1979. *Relative Slope Stability and Land Use Planning in the San Francisco Bay Region, California*. US Geological Survey Professional paper no. 944. Washington, DC: US Government Printing Office.

OAS. 1991. *Primer on natural hazard management in integrated regional development planning*. Washington, DC: Department for Regional Development and Environment, Executive Secretariat for Economic and Social Affairs, Organization of American States.

Sidle R.C., A.J. Pearce and C.L. O'Loughlin. 1985. *Hillside stability and land use*. Water Resources Monograph no. 11. Washington, DC: American Geophysical Society.

———. 1986. *Hillside stability and land use*. 2nd ed. Water Resources Monograph no. 11. Washington, DC: American Geophysical Society.

Thomas, D. 2004. Planning for hillside terrains. Paper presented at the workshop Enduring Geohazards (Landslides and Floods) in the Caribbean Region, Learning Resource Centre, University of the West Indies, St Augustine, Trinidad. 8 December.

CHAPTER 4

Developing a GIS-based Landslide Susceptibility Map for Tropical Mountainous Environments

SERWAN M.J. BABAN and KAMAL SANT

Abstract

Landslides are common phenomena on islands of the West Indies, due mainly to high rainfall and mountainous terrain. The landslide occurrence is frequently masked within events such as hurricanes or tropical storms, and it is therefore overlooked and not adequately taken into consideration in natural disaster preparedness and management at the national scale. Slope-stability models for mapping landslide susceptibility require a large quantum of parametric data about conditions that influence landslide occurrence. Frequently, this information is not available for wide areas since the cost of acquiring this information is prohibitive. Many developing countries, such as Trinidad and Tobago, suffer from a paucity of reliable and well-distributed data. The local absence of a landslide event recording system, coupled with a historically reactionary management approach, has limited the effectiveness of state agencies to efficiently undertake remedial works, with the sparse resources available. The use of GIS in landslide susceptibility mapping provides a valuable input into the decision-making process with respect to natural hazard management. This chapter describes the development of a GIS-based landslide susceptibility index and map for the island of Tobago, using the geo-environmental indicators of geology, slope, aspect, soil, rainfall, land use and a landslide inventory in a GIS environment. The susceptibility index for each slope is estimated. Slopes are classified into a landslide susceptibility range from severe, high, medium and low and are based on the density of landslide conditions within a management grid. The landslide susceptibility map presented is of value for general planning purposes, as well as for natural hazard mitigation and response.

4.1 Introduction

Natural hazard events, which include the occurrence of earthquakes, floods, volcanoes, tsunamis, windstorms and landslides, have constituted major disasters in all countries of the world. Globally, more than 300 disasters occur each year, taking about 250,000 lives and causing some US$60 billion in damage (Berz 1994). About 90% of impacts occur in developing countries, with a proportional loss of gross domestic product of twenty times that of developed nations (Alexander 1995). Population growth, and the resulting dispersal of settlements over hazardous areas, has served to increase the impact of natural hazards (Hansen 1984; Carrara and Guzzetti 1995). In recognition of this global problem, the United Nations declared the period from 1990 to 2000 as the International Decade for Disaster Reduction. This recognition fuelled the drive for research into the causes and effects of natural hazards. Regionally, the Caribbean Community (CARICOM), in conjunction with external agencies, has initiated and funded several programmes to mitigate the effects of natural disasters in member countries.

Of all natural hazards, landslides or mass movements are the most amenable to investigation and mitigation as they affect discrete areas and their processes are determinable. Landslides are phenomena that occur as a result of a number of determining and triggering factors, often with complex inter-relationships (Varnes 1978). Analysing landslides requires identification and modelling of the most important factors that contribute to slope failure. A landslide is defined as "the movement of a mass of rock, debris or earth down a slope" (Cruden 1991). The true impacts of landslides are often underestimated, as they frequently occur concurrently with other natural hazards such as earthquakes, floods and windstorm events and are therefore masked within the larger event (Johnson and DeGraff 1988; Ahmad and McCalpin 1999). The lack of landslide recording systems, coupled with the absence of centralized data collection for damage information, hampers efforts aimed at attempting to quantify losses and formulating policies and programmes, which can mitigate these negative effects (Baban and Sant 2004, 2005).

Landslides are caused by physical conditions at a location, which either leads to decreased ability of a slope to resist gravitational influence, an increased effectiveness of gravity acting on the slope or a combination of these two elements (Johnson and DeGraff 1988; Baban and Sant 2004). Several landslide classification schemes are in use. For example (Cruden and Varnes 1996), landslides are classified on the basis of the type of materials and movements involved. Terzaghi (1936) proposed a landslide classification system based on the physical properties of rocks. The use of geographic information systems (GIS) in the analysis of the spatial distribution of factors that contribute to landslide occurrence has grown in the last decade (Brabb 1984; Nagarajan et al. 1998; Barredo et al. 2000; Carrara et al. 2000). The GIS is uniquely capable of handling existing spatially diverse data sets from various sources, effecting the combination of these data sets, and highlighting and extracting spatial relationships (Baban and Sant 2004).

4.2 Study Area

The island of Tobago lies at latitude 11° north and longitude 60° west. It is approximately 41.5 km long, 12 km wide at its centre and has an area of approximately 292 km^2. The island forms part of the Republic of Trinidad and Tobago and is located 32 km northeast of the main island of Trinidad (Figure 4.1). Generally speaking, the northern side of the Main Ridge is steeper than the southern slope. Almost 80% of the island exceeds 152 m in elevation. The lowest "plain" area is located on

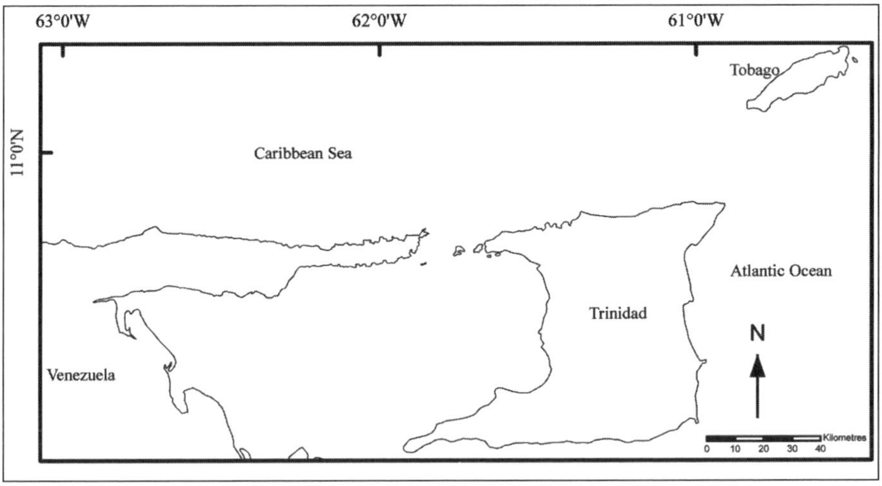

Figure 4.1 Location of study area.

the southwestern side of the island. This southwestern area is also the most developed (Baban and Sant 2004). Much of the interior of the island of Tobago is undeveloped due to the rugged terrain. This undeveloped area remains in natural forest and much of the Main Ridge area is a State Forest Reserve. Agricultural and residential developments have been limited mainly to a narrow coastal belt, and tourism has been intensively developed in the southwest parts of the island.

4.3 Methodology

4.3.1 Data Collection and Development

All landslide analyses require a landslide inventory. The Tobago House of Assembly and the Ministry of Works and Transport, on whom the joint responsibility for landslide response rests, were consulted. The reported absence of a landslide reporting system in Tobago necessitated the development of a landslide inventory from other sources.

Newspaper reports provided only two general locations of significant landslide events and approximate dates of occurrence of these events. The fact that the daily local newspapers were based in Trinidad, and not in Tobago, may have influenced the number of reports that could be located. Of those reports located regarding Tobago, the spatial description of the landslides could not be determined from the reports, but only the general areas for subsequent location/positioning during a field survey. Only recent landslides that occurred at Palatuvier Village and along the North Coast Road were identified from newspaper reports.

The National Emergency Management Authority (NEMA) Disaster and Response Hazard Map of 1995 is the only officially published landslide inventory for Trinidad and Tobago. The NEMA map was scanned, geo-referenced and landslide sites/areas digitized to yield landslide perimeter polygons. Each polygon was then assigned a unique identification number through which additional descriptive information such as landslide class, trigger(s), date of occurrence and other significant information could be recorded (Baban and Sant 2004, 2005). The classification scheme for landslides adopted for this paper follows Varnes's (1978) classification system. Although the spatial accuracy of the disaster and response map could not be ascertained, it was utilized as the only available official data source and was supplemented by a field surveyed landslide inventory to accommodate past and post-mapping period landslides.

4.3.2 Developing Landslide Geo-Environmental Based Indicators

Cruden and Varnes (1996) identified the most significant factors causing landslides in a tropical environment as comprising four main categories (Table 4.1). This chapter builds on this approach and advances the deductive concept for studying landslides based on the premise that conditions at known landslide sites within an area are reliable indicators of where future slope failures might occur. The favourable conditions to landslides in the study area, which are a combination of geo-environmental conditions, were determined based on the concept advanced by Cruden and Varnes (1996) and the authors' experience. The five geo-environmental indicators that contributed to landslide occurrence in Tobago were identified as *slope*, *aspect* (morphological), *geology*, *soil* (geological), *rainfall distribution* (physical) and *land use/cover* (human).

Table 4.1 Major Factors Causing Landslides in Tropical Environments

Cause Category	Description
Geological	Weak materials; sensitive materials; weathered materials; sheared materials; jointed or fissured materials; adverse oriented mass discontinuity (bedding); adverse oriented structural discontinuity (faults); contrasts in permeability; and contrasts in stiffness (stiff material over plastic material)
Morphological	tectonic or volcanic uplift; fluvial erosion; wave erosion of toe slope; erosion of lateral margins; subterranean erosion; deposition loading of slope or crest; and vegetation removal
Physical	Intense rainfall, prolonged exceptional precipitation; rapid draw down of floods and tides; earthquakes; volcanic eruptions; and shrink and swell weathering
Human	Excavation of slope toe or crest; loading of slope crest or toe; drawdown of reservoirs; deforestation; irrigation; mining; artificial vibration; and water leakage from utilities

Source: Cruden and Varnes 1996.

At known landslide locations, analyses need to be conducted to determine the specific geo-environmental conditions which represented the significant causes of landslide occurrence and which can collectively be used as landslide indicators, identifying and mapping likely landslides in Tobago. These indicators can then be used to predict the locations most likely to be affected by landslides, given similar conditions. The location and the spatial distribution of the landslide geo-environmental indicators (LGIs) throughout Tobago will be determined, via a vector-based GIS environment, through a quantitative combination of data themes representing each of the geo-environmental conditions in a vector-based GIS setting.

4.3.3 Developing Data Sets

Existing data sources located from which digital data sets were developed included the following:

1. The 1:150,000 scale NEMA 1995 Disaster and Response Map for Tobago published by the Lands and Surveys Division of Government of Trinidad and Tobago
2. Three 1:25,000 scale soil series maps for Tobago published by the Lands and Surveys Division of the Government of Trinidad and Tobago in 1974
3. Eighteen 1:10,000 scale topographic maps showing contours, rivers, roads and general land cover types published by the Lands and Surveys Division of Government of Trinidad and Tobago from 1962
4. Geological Map for Tobago (1:100,000 scale) published by the Ministry of Energy and Mines of Government of Trinidad and Tobago from 1998
5. Rainfall data from the Water and Sewerage Authority's Water Resource Agency published in the Environmental Management Agency's State of the Environment Report 1998
6. Three topographic maps (1:25,000 scale) published by the Lands and Surveys Division of Government of Trinidad and Tobago and dated as 1962
7. A watershed map developed by the Water and Sewerage Authority of Trinidad and Tobago published in the Environmental Management Agency's Sate of the Environment Report 1998

The process developed for extraction of data from these hard copy sources involved (a) the scanning of each map to produce a digital TIFF-format file; and (b) using Able Software Corporation's R2V (Version 5.5) to geo-reference; and (c) digitizing (vectorizing) each map; and (d) where necessary, joining data derived from adjacent maps. Once the raw digital vector files were derived, ESRI's Data Automation Toolkit was used to clean digitizing errors and build topology for each vector data layer. The built data layers were then imported into ESRI's Arc View (Version 3.2) and converted into shapefiles for attribute addition and spatial analysis. Each polygon had been automatically labelled with a unique reference number and relevant attribute textual information attached to the polygon's record attributes table. Table 4.2 provides a list of those data layers or themes developed and the relevant attributes or information attached. Primary data sets are those data

Table 4.2 Developed Primary* and Secondary** Data Layers and Attribute Information Recorded

Data Theme	Data	Type Attribute Information Recorded
*Soils	Polygon	Dominant soil type; secondary soil type; soil code
*Geology	Polygon	Geological formation name; formation period; geology code
*Rainfall	Polygon	Rainfall polygon map with mean annual precipitation
*Land Use/ Cover	Polygon	Land use/cover name; Land use/cover code
**Contours	Arc	Contour isoline value
**Rivers	Arc	River name and direction of flow
**Coastline	Polygon	Elevation value
**Watersheds	Polygon	Watershed name

derived directly from existing data sources, such as hard copy maps, in which possibly only a conversion to digital format is required, while secondary data are those derived from a combination of the conversion, manipulation and extraction of data from an existing data source. Consequently, primary data would contain a combination of the errors from the original data in addition to any errors introduced through data, conversion and attribute database table normalization processes. A secondary data set would contain a combination of the errors within the original data plus any errors introduced through data conversion, attribute database table normalization and all data manipulation required to produce the required data layer output.

Data verification comprises two phases involving the graphical and attribute parts of the GIS data layer. Phase one was determining whether any digitizing errors had been made by either not digitizing a polygon boundary or by digitizing a boundary that did not exist in the original data source. To detect these two common types of digitizing errors, both the geo-referenced TIFF scan of the original data and the developed vector data layer were displayed. The operator would then visually inspect each feature and verify the digitized features. Attribute

Table 4.3 Landslide-locations Derived Geo-environmental Indicators (Primary Datasets)

Geo-environmental	**Indicator Condition** (in order of prevalence)
Geology type	Parlatuvier formation (16); diorite – gabbro (6); unidentified volcanic and sedimentary rocks (17); ultramafic rocks (7); Argyle formation (12); Goldsborough formation (11); deformed mafic plutonic – volcanic (8); Bacolet formation (9); Mt Dillon formation (14); biotite tonalite (5); amphibolitic rocks (13); quaternary deposits (1); and Rockly Bay formation – Pliocene (4)
Rainfall	1,800 mm; 2,000 mm; 1,200 mm; 1,600 mm; 2,200 mm; 2,400 mm; 2,600 mm; 2,800 mm; and 1,200 mm (all ranges present)
Land use/cover	Low forest; high forest; shrub; urban; and agricultural (all ranges present)
Slope	20°–30°; 0°–10°; 10°–20°; 30°–40°; 40°–50°; 50°–60°; 60°–70°; and 70°–80°
Aspect	North (315°–45°); south (135°–225°); east (45°–135°); west (225°–315°); and flat (all ranges present)
Soil	Clays – 27, 43, 45, 5; clay loam – 13, 14, 16, 40, 42, 47, 52, 65, 70, 71; sandy clay loam – 15, 61, 62, 63, 64; and sandy loam – 24, 50

data was similarly added to the attribute database for the relevant data layer, and visually inspected to ensure compliance with the original data source and to evade polygons. The primary data sets comprise the data themes of geology, soil, rainfall distribution and land use/cover. Additionally, primary data themes were slope and aspect; however, these two themes had to be derived from a secondary set of data consisting of a TIN terrain model developed from the contours, rivers, coastline and watershed data layers (Table 4.3).

The topographic maps identified above were a significant data source as they provided the basis for the themes of coastline, rivers and contours. These secondary themes were used to develop a TIN for the island of Tobago. The TIN was verified by checking the elevations of known Lands and Surveys Division control stations against the extracted elevations from the control station planimetric positions on the TIN. The results indicated that the TIN was an average of 5 m above the surface terrain, as determined from the control stations. The TIN was exported as a three-dimensional polygon shapefile representing the terrain, and the slope and aspect for each TIN facet (polygon) was computed and attached to the polygon shapefile. To develop individual slope and aspect themes, the three-dimensional polygon shapefile was amalgamated to yield new boundary polygons with slope values within a specified 10° range (that is, re-classified). The same process was employed to separate the aspect data theme into the four cardinal directions and flat areas.

There is no recent land cover/use map for the island of Tobago. As a consequence, it was necessary to develop a provisional land cover/use map from the 1962 topographic maps. These maps contained inherent land cover/use interpretation, which was created from aerial photo interpretation of the source stereo imagery. This information was digitized and used to provide a land cover/use inventory at the time of the mapping. The original maps were scanned, geo-referenced, and the shading and land use/cover featured digitized. These features were then encoded with the relevant attribute and used to generate the provisional land-use map for the study area within the GIS. The GIS was used to generate Thiessen polygons around each point feature digitized, and similarly coded Thiessen polygons were subsequently dissolved to yield coded land-use polygon maps for the study area. Rainfall, soils and geology data layers were developed from published maps (Table 4.2) by a sequential process of scanning, geo-referencing, building topology and attaching the relevant polygon attribute information.

4.3.4 Identification and Mapping of Slopes Prone to Landslide Occurrence

The process for the identification of slopes prone to landslide was as follows (Figure 4.2):

1. *Development of the LGI:* The LGIs were developed using a sequential data theme combination process within the GIS, and provided a

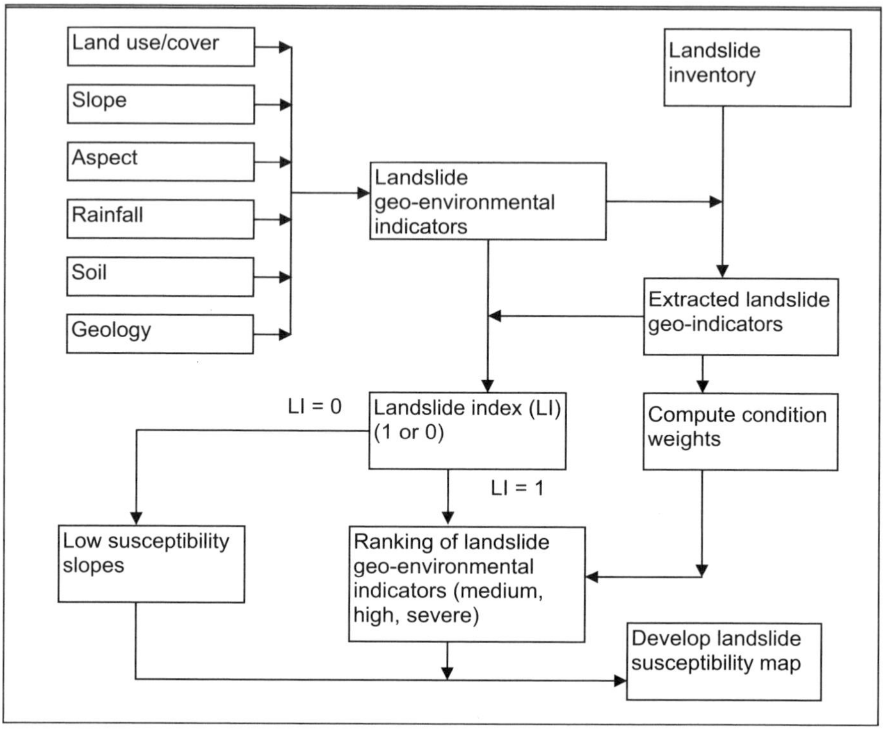

Figure 4.2 Development of landslide geo-environmental indicators and location of landslide prone conditions.

data theme that provided polygons with attribute values of slope, aspect, land use/cover, rainfall distribution, geology and soil over the study area.

2. *Implementation of the LGI:* Once the LGI was developed for the study area, those combinations of conditions that contributed to landslide occurrence were deduced by clipping the LGI with the landslide inventory polygon data theme. This process yielded those LGI conditions that existed at known landslide locations. The identified landslide-prone conditions were located within the LGI, and a new binary attribute, called the Landslide Index (LI), was added to the data theme, which would represent either the existence (value of 1) or presence/absence (value of 0) of a landslide-prone condition.

3. *Ranking of landslide-prone areas:* Slopes with landslide-prone condi-

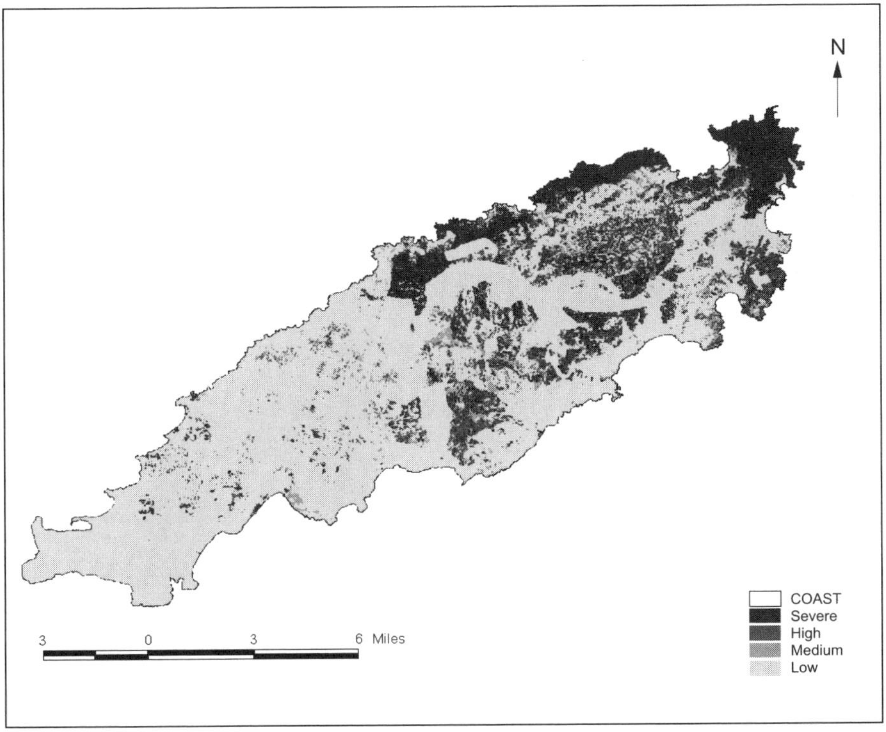

Figure 4.3 Developed landslide susceptibility distribution map for Tobago.

tions within the study area were those slopes with an LI attribute of 1, while slopes with non-landslide-prone conditions within the study area were those slopes with a LI attribute of 0. Slopes with a LI of 0 were labelled as having *low* landslide susceptibility. Slopes with a LI of 1 were weighted and ranked into three classes of medium, high and severe susceptibility to landslide occurrence (Figure 4.3).

4.3.5 Classification of Landslide-prone Slopes

The deduction of geo-environmental indicators for this study area, allowed the classification of slopes into landslide-prone and non-landslide-prone areas. The relative distribution of each environmental geo-indicator at landslide sites was also used to assign a ranking of slopes relative to future landslide occurrence by the assignment of weights to each geo-environmental indicator based on the prevalence of that condition, as determined by its planimetric area.

Weighting was carried out on the geo-environmental indicators; those indicators that were not found at known landslide sites were assigned a weighting value of 1, while those that were found at known landslide sites were assigned a weight (between 0 and 1) proportionate to the spatial extent or prevalence of the condition. Thus the study area's LGI was reclassified into a range of values to provide a susceptibility index (SI). The SI was then classified into four intervals representing severe, high, moderate and low susceptibility to future landslide occurrence.

4.4 Results and Discussions

The landslide inventory developed provided the spatial distribution and planimetric extent of 212 landslides (covering 44.64 ha), 51 of which were derived from the field-surveyed landslide inventory and the ortho-imagery. The landslide inventory represented approximately 1.5% of the total land space within the study area. The provisional land-use theme was classified into the five general land-use classes comprising agriculture (17.0%), high forest (22.5%), low forest (33.7%), shrub (18.0%) and urban (8.8%).

The initial LGI, developed for the study area and re-classified, indicated that 47.2% of the site possessed prone landslide geo-environmental indicators (that is, where the LI = 1). A summary of the deduced landslide-prone geo-environmental indicators is presented in Figure 4.4.

Weighting of each combination of geo-environmental indicators at landslide sites was affected by computing the total acreage of every unique combination of geo-environmental indicators and expressing each acreage as a percentage of the total area covered by landslides.

For example:

Area covered by a unique environmental geo-indicator combination = 0.25 ha

Total area covered by landslides = 44.64 ha

Weighted landslide susceptibility index = (0.25 / 44.64) × 100 = 0.056

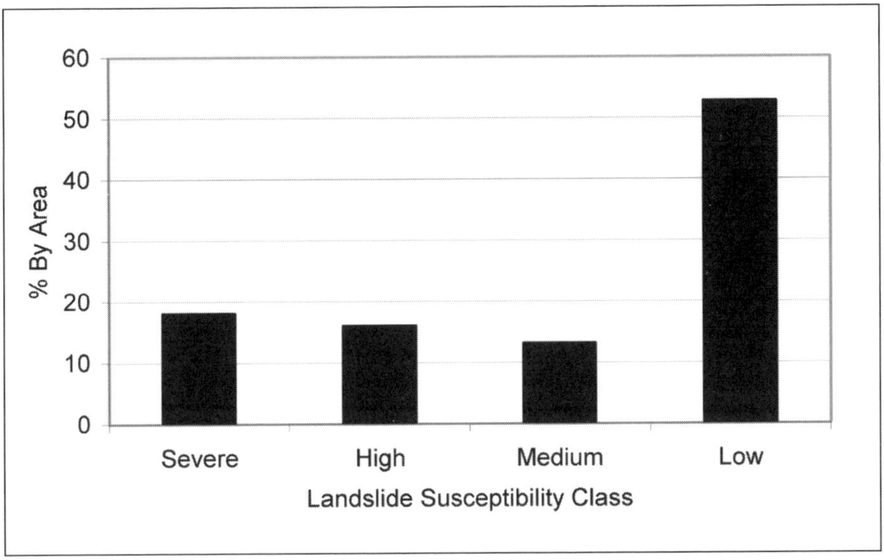

Figure 4.4 Environmental geo-environmental indicator landslide susceptibility classification for Tobago by area.

The landslide susceptibility index map, derived from the geo-environmental indicators at landslide sites, classified those areas with a LI of 1 into three equal ranges of severe (18.1%), high (16.0%) and medium (13.1%), and areas with a LI of 0 were classified as low (52.8%) (Figure 4.4). The spatial distribution of low susceptibility was concentrated to the southwest part of the island, with some bends stretching into the central portion of the Main Ridge. Severe susceptibility encompassed those areas (Figure 4.3) containing this combination of geo-environmental indicators deduced at landslide sites that had the largest physical or planimetric extent and were concentrated to the north and eastern sides of the island. Areas classified as having a high susceptibility rating were mainly located around the perimeter of the severely rated areas, while areas rated as medium susceptibility were located to the southeast and middle portions of the island.

The critical slopes map developed was compared to previous susceptibility maps developed by Baban and Sant (2004; 2005). There was congruity among all susceptibility maps in terms of the concentration of high landslide susceptibility on northwest-facing slopes along Main

Ridge as well with respect to the low susceptibility on the southwest region of Tobago. The even-weighted approach map was congruent with the critical slopes map for the central part of the island, but it showed a lower level of susceptibility when compared with the varied weighted susceptibility map. This may have been attributable to the additional landslide factors used in the analyses, as well as the differences in the spatial distribution and the quantum of landslides in the current landslide inventory.

4.5 Conclusions

The causal factors that were identified were primarily guided by the availability of data to represent the landslide causes in a tropical mountainous environment, as that which exists on the Caribbean island of Tobago. Additional data can assist in refining the approach, and the model is flexible enough to incorporate other landscape parameters within the analysis. The absence of a sufficient quantity of data with adequate spatial distribution of geotechnical soil parameters precluded the application of a numerical modelling approach for slope stability over a wide area in Tobago. Instead, a regional landslide susceptibility approach had to be implemented. The lack of a landslide reporting system limited the level of detail that could be examined by this analysis approach, as the types of landslides and period of occurrences represented within the NEMA map could not be ascertained.

The approach adopted assumed that each environmental geo-environmental indicator contributed equally to the occurrence of a future landslide event. This method does not predict when a landslide event will occur, but it instead provides the spatial location at which a landslide event may occur, since the conditions at such a site are the same as that at which a landslide has already occurred. This approach is not particularly suited for landslide hazard mapping, which also requires an indication of when the failure will occur, but is well suited to landslide susceptibility estimation. Hence, this approach supports the identification and ranking of critical slopes as a first-cut for future detailed geotechnical investigation and impacts upon the formulation of regional policies for developmental control. Furthermore, this is a macro scale approach and provides a rationale to decision makers in identifying landslide-prone slopes in Tobago. The classification of slopes presented can form the basis for prioritizing detailed geotechnical slope-stability investigations for mitigation purposes. This will enable the responsible authorities to move from a reactive mode into a proactive role, and reduce the effects of landslide occurrence in Tobago.

References

Ahmad, R., and J.P. McCalpin. 1999. *Landslide susceptibility maps for the Kingston Metropolitan Area, Jamaica, with notes on their use*. UDS Publication no. 5. Kingston: Unit for Disaster Studies, Department of Geology, University of the West Indies.

Alexander, D. 1995. A survey of the field of natural hazards and disaster studies. In *Geographical information systems in assessing natural hazards,* ed. A. Carrara and F. Guzzetti. Dordrecht: Kluwer Academic Publishers.

Baban, S.M.J., and K.J. Sant. 2004. Mapping landslide susceptibility on a small mountainous tropical island using GIS. *Asian Journal Geoinformatics* 5, no. 1:33–42.

———. 2005. Mapping landslide susceptibility for the Caribbean island of Tobago using GIS, multi-criteria evaluation techniques with a varied weighted approach. *Caribbean Journal of Earth Sciences* 38:11–20.

Barredo, J., A. Benavides, J. Hervas and C.J. van Westen. 2000. Comparing heuristic hazard assessment techniques using GIS in the Tirajana Basin, Gran Canaria Island, Spain. *International Journal of Applied Earth Observation and Geo-Information*, special issue: EU Runout Project, ed. C.J. VanWesten, 2, no. 1:9–23.

Berz, G. 1994. The insurance industry and IDNDR: Common interests and tasks. *Natural Hazards* 9, no. 3:323–32.

Brabb, E.E. 1984. Innovative approaches to landslide hazard risk mapping. *Proceedings of the Fourth International Symposium on Landslides*, 307–24. Canadian Geo-Technical Society, Toronto, Canada.

Carrara, A., M. Cardinali, and F. Guzzetti. 2000. Uncertainty in assessing landslide hazard and risk. *International Journal of Applied Earth Observation and Geo-information*, special issue: EU Runout Project, ed. C.J. VanWesten, 2, no. 1:172–83.

Carrara, A., and F. Guzzetti, eds. 1995. *Geographical information systems in assessing natural hazards*. Dordrecht: Kluwer Academic Publishers.

Cruden, D.M. 1991. Simple definitions of a landslide. *Bulletin of the International Association of Engineering Geology* 43:27–29.

Cruden, D.M., and D.J. Varnes. 1996. Landslide type and processes. In *Landslides: Investigation and mitigation,* ed. A.K. Turner and R.L. Schuester, 36–75. Transportation Research Board Special Report, no. 247. Washington, DC: National Academies Press.

Hansen, A. 1984. Landslide hazard analysis. In *Slope instability,* ed. D. Brunsden and D.B. Prior, 523–602. New York: John Wiley and Sons.

Johnson, R., and J.V. DeGraff. 1988. Landslide processes. In *Principles of engineering geology,* ed. R. Johnson and J.V. DeGraff, 439–56. New York: John Wiley and Sons.

Nagarajan, R., M. Anupam, A. Roy and M.V. Khire. 1998. Temporal remote sensing and GIS application in landslide hazard zonation of part of Western Ghat, India. *International Journal of Remote Sensing* 19, no. 4:573–85.

Terzaghi, K. 1936. Stability of slopes of natural clay. In *Proceedings of the First International Conference on Soil Mechanics,* vol. 1, 161–65. Harvard University.

Varnes, D.J. 1978. Slope movement type and processes. In *Landslides: analysis and control,* ed. R.L. Schuster and R.J. Krizek, 20–47. Transportation Research Board Special Report, no. 176. Washington, DC: National Research Council.

CHAPTER 5

Using Contemporary Geo-imaging Technologies for Landslide Investigations in Tropical Environments

RAID AL-TAHIR and VERNON SINGHROY

Abstract

Landslide hazards occur in many places around the world and pose serious threats to settlements, infrastructure, transportation, natural resources and tourism. Each year, these hazards cost billions of dollars and cause numerous fatalities and injuries. Landslide identification and mapping are essential for landslide risk and hazard assessment. Because they are highly dynamic events and activities, there is also a need for multitemporal monitoring of landslides for the knowledge and the prediction of their possible spatial and temporal evolution. Such information is essential for informed decision making by scientific and resource management authorities to detail the threats as well as to establish safeguard measures. However, major obstacles in this endeavour include lack of data and understanding of factors controlling the processes involved. Remote sensing technologies have great potential in overcoming the information void in the Caribbean region. They are relatively inexpensive and have the ability to provide information on several parameters that are crucial to landslide identification, mapping and monitoring. This information can be directly integrated into GIS for analysis and decision making at an economic cost/benefit ratio.

This chapter argues that the gap in data and information can be managed through the adoption of remote sensing technology. It surveys the current progress and innovative trends in this field. It also examines the use of the new imagery data as an up-to-date and affordable source of information for establishing the necessary baseline information for landslide management in the Caribbean region.

5.1 Introduction

Landslides are defined as the movement of a mass rock, debris or earth down a slope (Cruden 1991). Landslide activity worldwide is increasing and accounting for enormous annual property damage in terms of both direct and indirect costs. This trend is expected to continue because of the increased encroachment of developments into hazardous areas, expansion of transportation infrastructure, deforestation of landslide-prone areas and changing climate patterns (Dai et al. 2002; Schuster 1996; Spiker and Gori 2003). The increasing impact of landslide hazards can be curbed through better understanding and mapping of the hazards and improved capabilities to mitigate and respond to the hazards. Successful landslides management, though, must account for a wide range of parameters and data. The set of required physical information includes topography and terrain, soil types, watershed/catchments, land cover and forestry, and the intensity of the triggering factors (Soeters and van Westen 1996; van Westen 2004). Ultimately, landslides risk management requires socioeconomic data (housing location, valuation data, demographic structure, census information) as well as land-use information, administrative boundaries, development pressure, land-use capability and environmental constraints.

However, there is a severe general shortage of reliable and compatible data sets in the whole Caribbean region. Information needed for accurate planning is often outdated, non-existent, or expensive and time-consuming to collect (Al-Tahir et al. 2007). Without such information, the investigation of landslide susceptibility and the formation of proper national planning policies in many Caribbean island states are both difficult and error-prone (Baban and Sant 2005). The nascent remote sensing technologies provide an excellent source for collecting primary geo-data because landslides directly affect the ground surface. Additional information can be extracted from remote sensing images about terrain conditions that are critical in determining a site's susceptibility to slop instability (Soeters and van Westen 1996).

Considerable advances in remote sensing technology have occurred both in acquiring digital aerial photography and high-resolution satellite data. Parallel to this, new techniques have been developed for improved processing and extraction of spatial information from these new data sets. Recent improvements in computer software and hardware have allowed remote sensing and geographic information systems (GIS) to provide the way forward to collect and manage relevant data sets, and development of management scenarios to evaluate mitigation strategies (Ehlers 2004; van Westen 2004).

This chapter examines current progress and trends in remote sensing technology. It presents ways where remote sensing can provide a suitable alternative to collect spatial data necessary for effective landslide investigations in the Caribbean. Section two looks specifically at the latest developments in remote sensing, while section three discusses the general directions in the use of aerial photograph and satellite remote sensing technologies in the studies of landslide. Interrelated to the section's theme, two specific case studies will also be presented. Conclusions are then presented in the last section.

5.2 Advancements in Geo-imaging Technology

Remote sensing of the environment involves the measurement of electromagnetic radiation reflected from or emitted by the Earth's surface and relating these measurements to the types of land cover and habitat in the area being observed by the sensor. Photogrammetry has often referred to techniques handling aerial or terrestrial images, while remote sensing dealt with satellite imagery. This simple separation between photogrammetry and remote sensing was probably based on the fact that each of them provides some capabilities that cannot be achieved by the other. Among others, the comparative capabilities include ground coverage, repeatability of observations, spectral ranges and geometry for three-dimensional mapping (Li 1998). The following subsections detail the advancements in the technologies of photogrammetry and satellite remote sensing.

5.2.1 Digital Photogrammetry

The field of photogrammetry is rapidly changing with new technologies and protocols being developed constantly. In a relatively short period of time, the practice of photogrammetry has gone from analog to digital (or softcopy) with the advent of computing and imaging technology. The main driving premise in developing digital photogrammetry has been that it would enhance the performance, speed and accuracy in the execution of photogrammetric tasks (Crystal 2003). Progress has occurred along two tracks: developing commercial digital cameras for direct capturing of digital images and developing digital photogrammetry systems for data processing and information extraction.

Digital Aerial Cameras

The most obvious requirements for digital photogrammetry are the digital images themselves. While these may be obtained by scanning aerial photographs, the emerging trend is the use of digital airborne cameras. Direct digital photography is capable of delivering photogrammetric accuracy and coverage as well as multispectral data at any user-defined resolution up to 0.1 m ground sampling distance (Keating et al. 2003). The new digital cameras combine photogrammetric positional accuracy with multispectral capabilities for image analysis and interpretation. Coupled with differential GPS and inertial navigation systems (INS), these sensors generate geo-referenced, ultra high-resolution multispectral image data.

As there is no chemical film processing, the direct digital acquisition can provide image data in just a few hours after the mission is flown, compared to several weeks using the traditional film-based camera (Keating et al. 2003). Another advantage over traditional film is the ability to assess the quality of data taken directly after the flight is completed. Additional advantages of digital cameras are better radiometric image quality (due to direct sensing), "non-ageing" storage, and direct integration into GIS and image processing systems (Ehlers 2004). Two digital mapping cameras, ADS40 by Leica Geosystems and DMC from Z/I Imaging, were first presented to the market in 2002 to address requirements for extensive coverage, high geometric and radiometric resolution and accuracy, multispectral imagery, and stereo capability. These two cameras and the successive ones from other companies (for example, DiMAC [www.dimacsystems.com], DSS by Applanix [www.applanix.com], JAS150 from Jena-Optronik GmbH [www.jena-optronik.de], and Vexcel's UltraCamD and UltraCamX [www.vexcel.com]) are generally either based on one-dimensional linear or two-dimensional arrays technologies of CCD sensors to accomplish an airborne digital recording system.

The Leica Geosystems Airborne Digital Sensor (ADS40) utilizes triplet linear arrays to implement the three-line-scanner concept. This concept generates one image looking forward, another one looking vertically down and a third one looking backward from the aircraft (Figure 5.1a). The ADS40 simultaneously captures data from three panchromatic as well as four multispectral bands that receive information from exactly the same portion of the Earth's surface through a special beam splitter and filter. These concepts have the benefits of reducing the

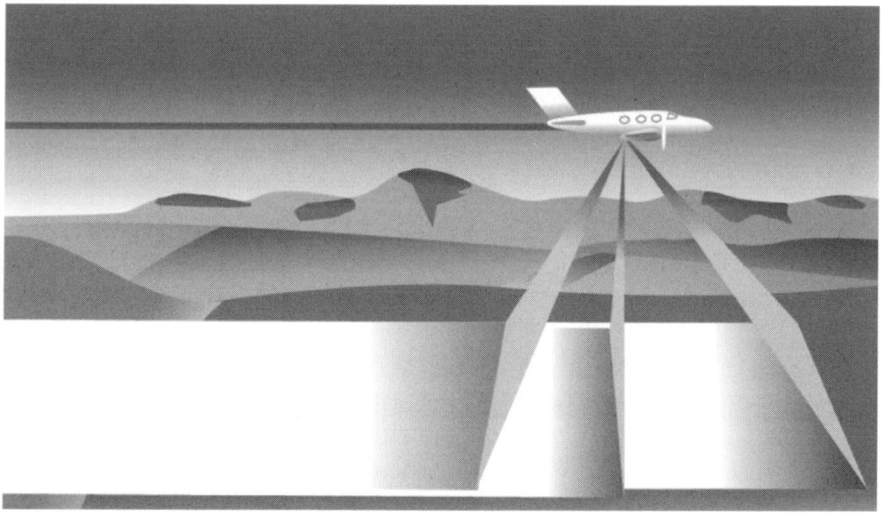

Figure 5.1a The three-line scanning principle in ADS40 (Leica 2002).

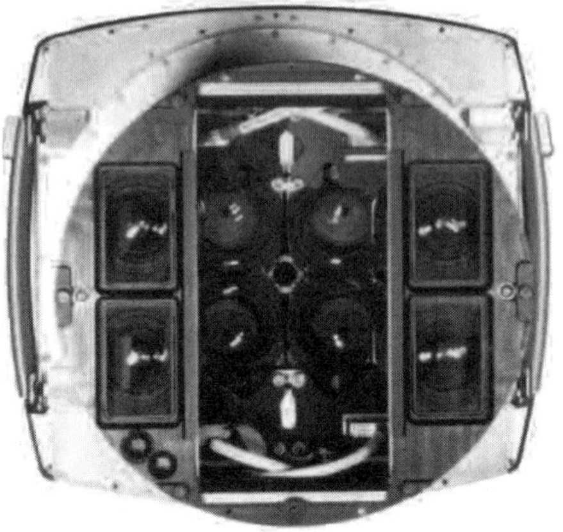

Figure 5.1b The arrangement of multiple CCD cameras in DMC (Z/I Imaging 2005).

ground control requirements, producing high-quality digital terrain models (DTM) and a perfect RGB co-registration (Leica 2002). However, on the downside, the airborne line-scanning system requires incorporating inertial navigation system and real-time kinematic GPS positioning to rectify each line and improve the geometric accuracy of the final scene. The second camera, the Digital Modular Camera (DMC), developed by Z/I Imaging, makes use of two-dimensional arrays and a set of coupled nadir-looking lenses to emulate a standard frame camera's central perspective (Hinz et al. 2001). The DMC's recording system comprises of up to eight individual, yet synchronously operating, CCD array cameras that can be put together in a modular design (Figure 5.1b). The high-resolution panchromatic channel contains four converging 7 k × 4 k large area chips and high-performance lenses that provide a single, digitally mosaicked image of 7,680 pixels along track and 13,824 pixels across track. For the simultaneous collection of true and false colour images, four multispectral channels are incorporated in the camera electronics unit, each of which features a separate wide-angle lens with a 3 k × 2 k CCD chip (Z/I Imaging 2005).

Softcopy Workstations

Digital photogrammetric workstations (DPW) are used to process digital images (aerial and satellite imagery) and are on the verge of replacing the current photogrammetric instruments. A DPW consists of hardware and software components that accept digital photographs/images, interactively and/or automatically perform photogrammetric procedures and operations, and produce digital and paper outputs. Typically, a DPW consists of a graphics workstation with a stereo viewing device and a three-dimensional mouse (Tao 2002). Digital stereo plotters are around three to four times cheaper than analytical stereo plotters (Crystal 2003). At present, high-end DPWs support automatic or semi-automatic processing of specific functions that are otherwise extremely labour intensive. These include Geomatica by PCI Geomatics (www.pcigeomatics.com), LPS by Leica (gi.leica-geosystems.com), ImageStation from Intergraph (imgs.intergraph.com), SOCET SET by BAE Systems (www.socetgxp.com), ER Mapper (www.ermapper.com) and Summit Evolution from DAT/EM Systems International (www.datem.com).

The use of digital images permits vastly extended automation possibilities that enable quick and efficient production of digital terrain models (DTM), ortho-rectified images and extracted vector features. The generation of DTM is practically done automatically through image matching that identifies and measures corresponding points in two or more overlapped photographs or images (Tao 2002). A similar degree of automation has also been achieved in producing ortho-images. Ortho-images have been one of the driving forces in the adoption of DPWs as they are a preferable product for many GIS applications since features can be delineated on top of ortho-images without stereo viewing (Keating et al. 2003). However, automation in the field of feature extraction from imagery is still limited, despite it being one of most important tasks in photogrammetry. Some vendors provide semi-automated tools to help the manual process, but the performance of such tools still needs improvements in terms of reliability. Notwithstanding, significant research efforts have been devoted through adapting higher-level image processing and image understanding techniques (Tao 2002).

5.2.2 High-Resolution Satellite Remote Sensing

Remote sensing–based data collection and research for the environment has been predominantly founded on using mid-resolution satellite imagery. Three platforms are currently in orbit and obtaining data: the US Landsat, the French Spot and the Indian IRS programmes. All three systems have a swath width of 60–180 km and produce multispectral

Table 5.1 Satellite Parameters and Spectral Bands (Digital Globe 2003; Orbimage 2003; Space Imaging 2003)

		Ikonos	**QuickBird**	**OrbView-3**
Sponsor		Space Imaging	Digital Globe	Orbimage
Launched		September 1999	October 2001	June 2003
Spatial resolution (m)	Panchromatic	1.0	0.61	1.0
	Multi-spectral	4.0	2.44	4.0
Spectral range (nm)	Panchromatic	525–928	450–900	450–900
	Blue	450–520	450–520	450–520
	Green	510–600	520–600	520–600
	Red	630–690	630–690	625–695
	Near Infrared	760–850	760–890	760–900
Swath width (km)		11.3	16.5	8
Off nadir pointing		26°	30°	45°
Revisit time (days)		2.3–3.4	1–3.5	1.5–3
Orbital altitude (km)		681	450	470

data (visible and near infrared) and short-wave infrared (SWIR) with a ground resolution of 10 m to 30 m. All of these instruments have been built and operated through government-sponsored programmes.

Since the late 1990s, private satellite corporations started collecting high-resolution remote sensing data. The satellites from Space Imaging (Ikonos), Digital Globe (QuickBird) and Orbimage (Orbview-3) are already in orbit, capturing imagery at up to 0.61 m ground resolution. These systems share several common specifications with respect to the spectral and spatial resolutions as well as orbital details. Table 5.1 lists selected information about the satellite systems being discussed, including data about ground resolution, spectral coverage and swath width.

The new satellite images are recorded with 11–bit dynamic range, extending the pixel values to 2048 gray shades. Practically, it means that greater detail can be extracted from scenes that are very dark (for example, shadows) or very washed out from excessive sun reflectance (Corbley 2000). Additionally, 1 m colour imagery can be created using a pan-sharpening process that combines the high spatial resolution of the panchromatic image with the spectral information of the multispectral bands. Digital Globe and GeoEye (a merger of SpaceImaging and Orbmage) have initiated plans for their next generation systems (WorldView and GeoEye-1, respectively) to be launched during 2007 and 2008. The new systems will have enhanced collection capacities and revisit capabilities and will have a better than 0.5 m resolution. Consequently, the end users will soon have access to images and information of higher resolution.

The new high-resolution sensors pose new challenges for automated interpretation, extraction and integration of information. Finding features in sub-metre imagery is a new challenge since most feature extraction techniques have been developed for lower resolutions. It is therefore essential that new techniques be developed that allow automated processing of high-resolution and multisensor images as well as accurate interpretation results. One of the promising approaches is the use of auxiliary spatial (contextual) information besides the multispectral information in the processing and classification steps (Ehlers 2004).

5.2.3 Radar (SAR) Remote Sensing

The high-resolution images that are provided by RADARSAT-1 (8 m) RADARSAT- 2 (3 m) TerraSAR × (1 m) and ALOS (10 m) are especially useful for landslide inventory and mapping landslide geomorphology. Figure 5.2 shows current and future SAR missions that have the capability for landslide inventory and monitoring.

RADARSAT 2, which was launched in March 2007, has several capabilities that will be useful to geologists. Some of these capabilities include the availability of high-resolution 3 m SAR images, multi-polarization and fully polarimetric image modes, and left and right looking images (Morena et al. 2004). The simultaneous right and left looking capabilities of RADARSAT-2 (Figure 5.3) are particularly useful to

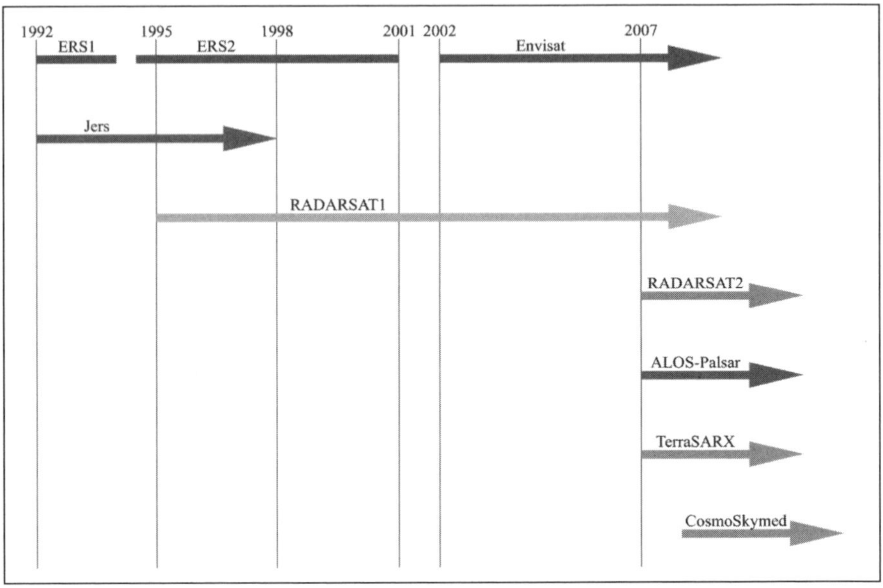

Figure 5.2 Future radar satellites.

monitor landslide process along N-S valley slopes in the line of site of the satellite and decrease the revisit time for greater monitoring efficiencies.

These enhancements are of high relevance for landslide hazard assessment and monitoring. The ultra-fine beam improves object detection and classification, while the multi-polarization mode produces better discrimination of various surface types and improved object detection and recognition (MDA 2006).

InSAR stands for Interferometric Synthetic Aperture Radar. InSAR is a proven technique for mapping ground deformation using radar satellites. It has been used in monitoring motion from earthquakes, volcanic activity, landslides and subsidence. InSAR greatly extends the ability of scientists to monitor landslides because, unlike other techniques that rely on measurements at a few points, InSAR produces a spatially complete map of ground deformation with centimetre-scale accuracy without subjecting field crews to hazardous conditions on the ground. An interferometric image represents the phase difference between the reflected signals in two SAR images obtained from similar positions in

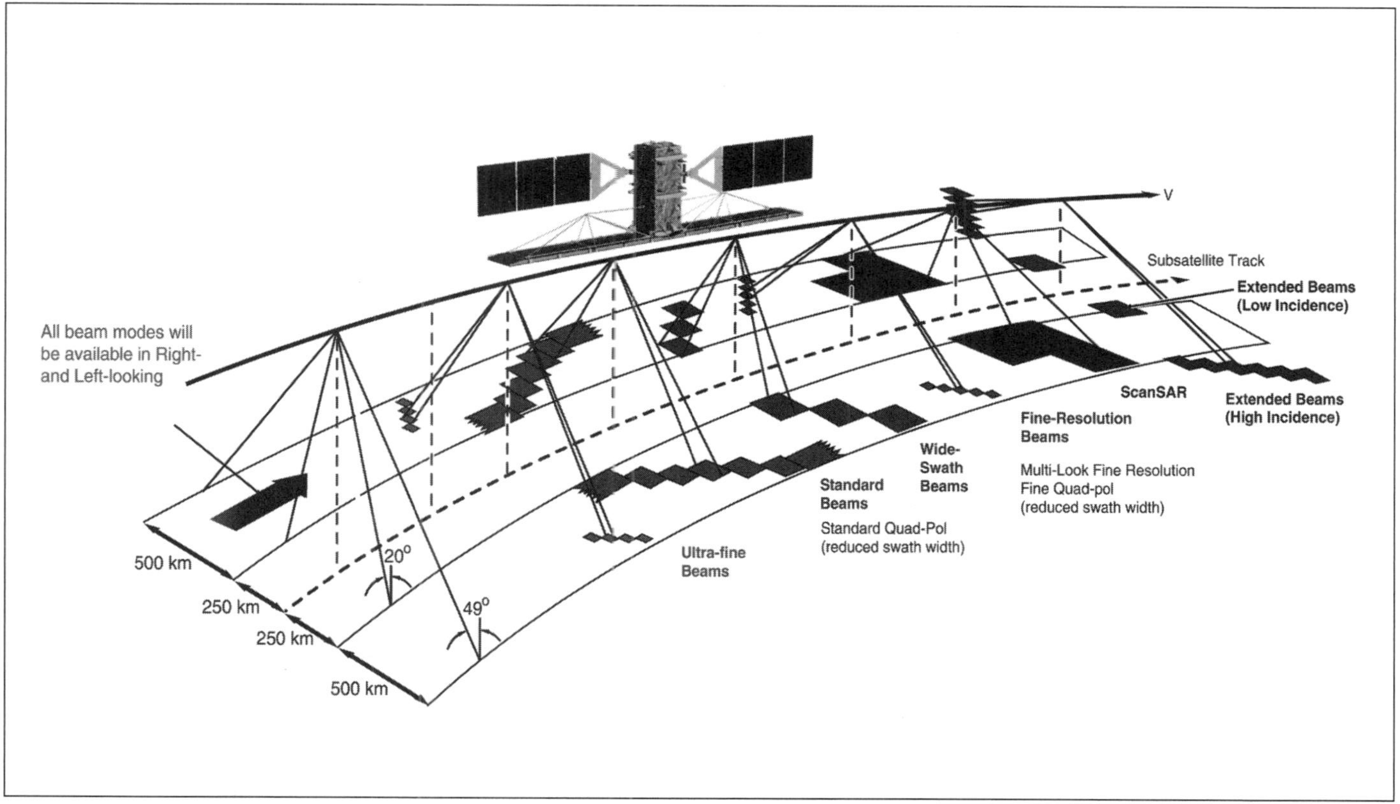

Figure 5.3 Viewing geometry for RADARSAT 2 (Morena et al. 2004).

space. In case of space borne SAR, the images are acquired from repeat pass orbits. For RADARSAT, the standard orbital repeat interval is 24 days.

5.3 Applications of Geo-Imaging in Landslide Studies

Remote sensing data are often used in the three main stages of a landslide-related investigation: detection and identification of landslides, monitoring of existing landslide, and spatial and temporal analysis and hazard prediction (Metternicht et al. 2005). At the first stage, it is required to view the size and contrast of the landslide features and the morphological characteristics of the topography within and around the landslide. Parameters of interest are the type of movement that has occurred, the degree of present activity of the landslide and the depth to which movement has occurred. The second stage in a landslide study is typically concerned with monitoring the movement of a landslide to assess its activity. This involves the comparison of landslide conditions over time, including the extent of the landslide, the speed of movement and the change in the surface topography (Metternicht et al. 2005). The third phase aims at predicting location of next likely failures to provide landslide hazard information needed for planning and protection purposes. Landslide hazard can be normally predicted based on the assumption that landslides are most likely to occur in conditions similar to those that have caused past failures (Soeters and van Westen 1996). Hence, the knowledge of the location, type and distribution of landslides occurring over time is essential for forecasting the future evolution of the landslide in an area.

Satellite remote sensing in the optical region of the electromagnetic spectrum has been scarcely used for direct landslide studies mainly due to insufficient spatial resolution by most space borne earth observation systems (Hervás et al. 2003; Soeters and van Westen 1996). Optical-IR satellites are applied, instead, to the mapping of landslide-related factors that fall more within the environmental and human categories (for example, land-use and geological details) that assist in analysing the relationships between landslides and their causative factors (Metternicht et al. 2005). Meanwhile, aerial photographs have become a standard and indispensable tool in the study of landslides because of the diagnostic morphology created by some mass movements (for example, disrupted vegetation cover, scarps). The advantage of aerial photographs in such a task stems from the fact that they provide high spatial resolution and synoptic view of an entire area that allows the user to see features, patterns and trends that cannot be seen on the ground. Moreover, they can be repeated at different time intervals permitting multi-temporal analysis (Ciciarelli 1991; Karsli et al. 2004).

Using photo interpretation techniques, efforts have mainly concentrated on extracting possible indirect landslide indicators such as land cover disruption patterns. Stereo photogrammetry technique is particularly worthwhile as stereo models depict, three dimensionally, the typical morphologic features of the landslides and the state of the surrounding vegetation. This can provide diagnostic information that can reveal the type of slide, depth, vegetation and drainage conditions of the landslide. Another use for stereo photogrammetry is in generating digital elevation models essential for detecting and monitoring landslides, especially the smaller scale slides. Long-term landform evolution of landslides can be measured from multi-temporal digital elevation models derived from sequential photo stereo pairs (Hervás et al. 2003).

However, the recent advent of high spatial resolution satellite imagery has opened new perspectives for detecting, monitoring and predicting landslides (Hervás et al. 2003; Metternicht et al. 2005). Increased detail adds an entirely new level of geographic knowledge to image-based spatial information and GIS databases. The less than 1 m ground spatial resolution allows users to identify and map small objects that were previously not detected in the coarser satellite imagery (Li 1998). The new high-resolution aerial and satellite sensors are now capable of capturing data that would be suitable for mapping at scales of 1:5,000 or better, as compared to 1:50,000 scale mapping from existing mid-resolution satellites.

The improved spatial characteristics have also influenced the accuracy of the extracted information. With the aid of ground control points for referencing the images, the spatial accuracy can further be improved to 2 m horizontal accuracy and 3 m vertical accuracy; this is equivalent to 1:2,500 scale map standards (Corbley 2000). Consequently, large scale digital mapping products, such as digital elevation models and digital orthophoto and line maps can be more accurately produced (Li 1998) and used for more reliable assessment of factors related to landslide as well as recovery efforts. Added to the advantages, the revisit rate of 1 to 4 days for the new high-resolution satellites significantly improves their temporal resolution compared to the 16 to 21 days' rate of the earlier mid-resolution systems. It is possible now to frequently map an area without special flight planning and thus scheduling as required in aerial photogrammetric data acquisition (Li 1998), and provide adequate continuous monitoring of ongoing events or responding to landslide disasters.

On the parallel track, the advancements in digital photogrammetry discussed in the previous section have transferred the photogrammetric workflow into digital. This has direct impact on the relevance and effectiveness of using aerial photography in investigating landslides. Because of their digital nature, aerial images now are acquired in near-realtime. The higher radiometric capabilities (11-bits imaging) mean better identification of features, even in badly illuminated locations and shadows. These images can be rapidly converted into accurate digital elevation models and orthophoto maps owing to the greatly automated processing of the highly developed photogrammetric software suites. This means more details in higher accuracy at a short time, all of which is essential for reliable and effective study of landslides, especially for monitoring and response planning during a disaster.

This section has synthesized the different approaches and the foreseen potentials for using satellite and aerial images. One may consult Mantovani et al. (1996), Metternicht et al. (2005) and Singhroy (2002) for more comprehensive reviews and case studies. In the following sections, two specific case studies will be looked at. The first one uses SAR technology for investigating rockslide avalanche, while the second uses stereo analysis of aerial photographs to develop landslide inventory.

5.3.1 Radar Application in Landslide Investigation and Visualization

Remote sensing techniques are increasingly being used in slope-stability assessment (Alberta Environment 2000; Murphy and Inkpen 1996;

Singhroy et al. 1998). Differential synthetic aperture radar (SAR) interferometry has been shown to be capable of measuring landslide displacement fields of centimetre order over relatively large areas and monitoring landslide pre- and post-slide activity over different geological/ topographic and triggering mechanisms (Hervás et al. 2003; Rott et al. 1999). However, for this technique, slope activity is to be monitored under specific conditions, such as InSAR coherence over long periods, data pairs with short perpendicular baselines, short time intervals between acquisitions and correcting the effect of topography on the differential interferogram.

The Frank Slide, a 30×10^6 m^3 rockslide avalanche of Paleozoic limestone, occurred in April 1903 on the east face of Turtle Mountain in southern Alberta, Canada. Seventy fatalities were recorded. This slide is still active. Several investigations have focused on characterizing grain size and distribution of this rock avalanche in order to understand post failure mechanism and mobility (Singhroy and Mattar 2000). Factors contributing to the slide have been identified as the geological structure of the mountain, subsidence from coal mining at the toe of the mountain, blast induced seismicity, above-average precipitation in years prior to the slide and freeze-thaw cycles (Singhroy and Molch 2004). In 2001, 6,000 tons of rock fell from the north slope of the Frank Slide that led to this InSAR investigation. The Government of Alberta has installed GPS stations and several in situ monitors to monitor post-slide activity at specific locations. In this study, InSAR results assist in locating in situ monitors, as well as provide a regional and seasonal view of gravitational mass movement. For the Frank Slide, coherence values are generally high on the slide itself, even for temporal baselines of more than 700 days. This can be attributed primarily to the lack of vegetation on the slide and indicates a general stability of the individual scatterers on the slope.

The post-failure mechanism and mobility of the Frank Slide InSAR images (Figure 5.4) are related to seasonal and moisture conditions. For instance, after heavy rainfall during the October to November 2003 time period, the localized slope deformation is the result of gravitational mass movement and local surficial slope failure within the colluvium (for example, old and recent rock fall debris) accumulated at the base of the slope. During the spring months (for example, April 2004)

Figure 5.4 Frank Slide InSAR images.

the availability of extensive moisture from snow melt triggers active surficial motion processes, resulting in numerous zones of significant vertical surface displacement, both positive and negative. In the springtime, the surficial processes increase significantly. The deformation at the base of the mountain slope is related to settlement of colluvium and rock avalanche debris.

SAR visualization techniques using a combination of digital elevation models (DEM) and SAR images are useful three-dimensional images for interpretation regional slope morphology and can be considered useful first steps for regional landslide inventory and monitoring. In the case of landslides in the Canadian Rockies, such visualization was used to interpret fault lines and slope morphology of large landslides and land use/cover. These parameters combined in one three-dimensional image can provide an effective interpretation of areas of potential landslides in seismically active areas or areas where excessive rainfall may trigger landslides and mudflows.

Figure 5.5 provides an example of a regional three-dimensional combined SAR/DEM of the North Range in Trinidad. In most tropical areas where there is usually a lack of cloud-free optical images, the radar images provide the pseudo-stereo geomorphological image as well as

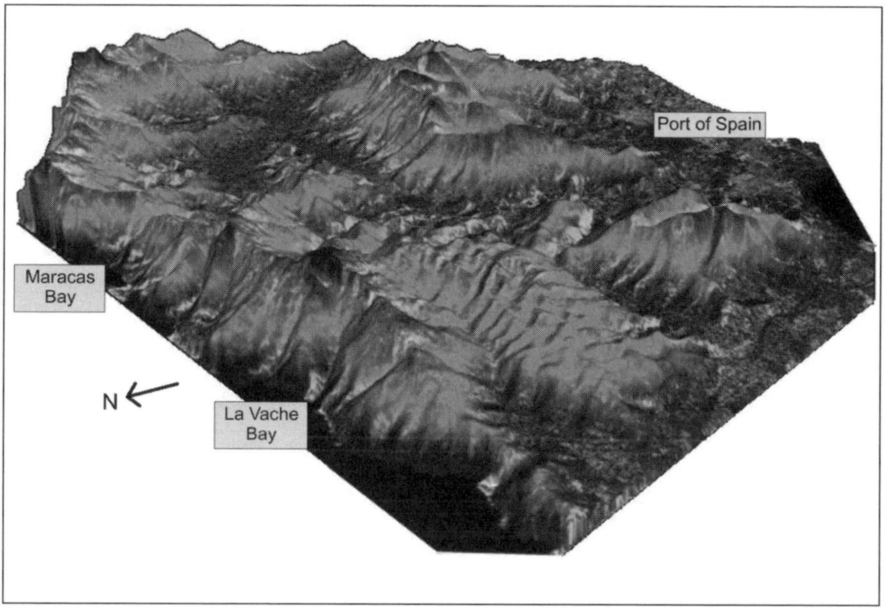

Figure 5.5 RADARSAT/DEM perspective view of Port of Spain – Maracas Bay.

land use/cover and the distribution of populated and infrastructure information. The DEM assists in correcting for layover and shadowing on the SAR image. These visualization images are a useful first step in regional landslide inventory and monitoring in relationship to topography and land use/cover in populated areas.

5.3.2 The Use of Aerial Photographs for Landslide Inventory

The objective of this study was to provide an inventory of historical landslides in the western side of the Northern Range in Trinidad. Such information is the base for other landslide hazard techniques (Soeters and van Westen 1996). The methodology adopted in this study relies on using aerial stereo photogrammetry for the detection of landslides and the quantification of their physical characteristics.

Scarps of historical landslide may not be detectable on aerial photographs as they are most likely covered by vegetation. This is most definitely the case in a tropical environment. Likewise, other geological clues, such as rocks, bedrock and unconsolidated material and geological structure, may not be evident as well. Landslides must then be inferred using the elements of photo interpretation to identify some diagnostic patterns and indicators of landslides based on morphology, vegetation and drainage patterns that are seen in the stereo model. Several of these indicators are discussed in Ciciarelli (1991) as well as Soeters and van Westen (1996). However, in this case, one must adopt the most relevant and significant of these in relation to the tropical mountainous environment (Al-Tahir and Thompson 2007).

The first phase in this process is thus to identify and acquire relevant aerial photographs. Forty-one photographs at scale of 1:25000 in one strip were obtained covering the study area, thus creating 40 stereo models. These photographs were part of the comprehensive mission of aerial photography in 1994 that was used for the production of the national base map. The photographs were then scanned at 800 dpi resolution for input into the softcopy photogrammetric system. The second phase in the methodology is the orientation of the stereo models, which is vital for establishing the true geographic position, scale and tilt of the stereo model. By the end of this stage, each ray from one photograph will intersect with the corresponding ray from the other photograph, creating the three-dimensional model in the geographic frame of reference. This phase depends on having control points with ground (map) coordinates to properly scale and level the stereo model. Considering the photo scale and the difficulty gaining access to the area, the use of maps at a scale of 1:25,000 was deemed sufficient for providing control for this task. The last phase is concerned with the collection of significant information related to landslide forms through the use of stereo photogrammetry. Following the photo interpretation principles, the inspection of landslides drew on identifying the concave upslope source and convex downslope deposit, as well as inspecting the main scarp. Additionally, the vegetation and drainage of these historical landslides were examined. Overall, the concavity coincided with tonal differences in the vegetation as shown in Figure 5.6. Photo interpretation gave the interpreter an appreciation of the terrain surrounding landslide sites based on wider coverage. Aspects of the environment, soil characteristics, vegetation, morphology and drainage conditions are of significant importance in this respect.

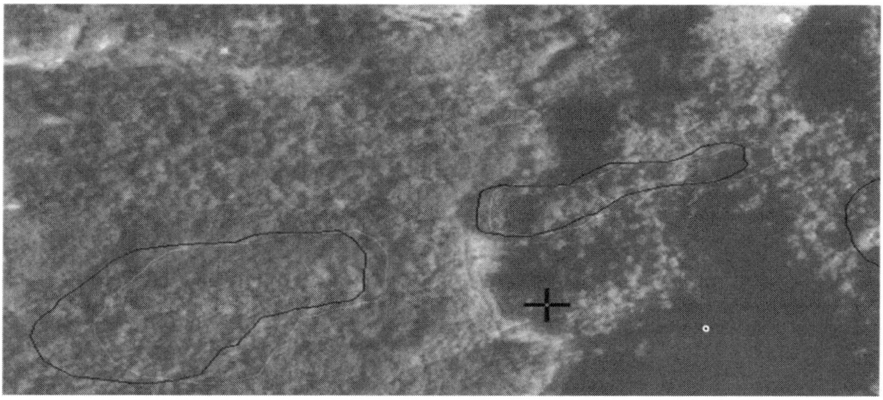

Figure 5.6 Two landslides as depicted on the photographs.

The spatial characteristics of detected landslides were extracted by stereo photogrammetry and brought into GIS software, where attribute data were added to each landslide. The final maps portrayed the distribution and geographic location of the historical landslides detected within the study (Figure 5.7). By the end of this study, a total of 40 stereo models were created and inspected. Eleven landslide forms were detected: six translational landslides, four rotational landslides and one earth flow slide.

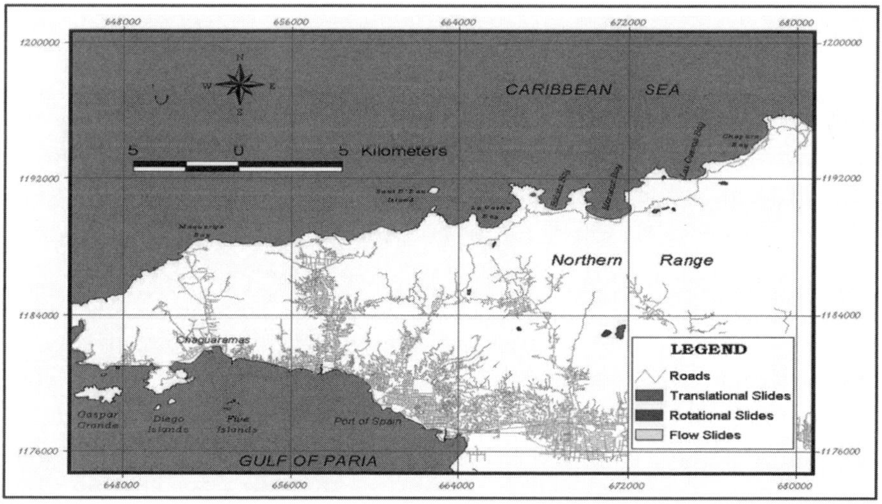

Figure 5.7 Location of landslides detected in the study area.

Based on the vegetation coverage and the morphology, all detected landslides appear to be of a historical nature rather than a recent occurrence. Some of the detected landslides are close to roads as seen in Figure 5.7, while the rest fall far away from the road network. This is a plus point for the method as it could identify slides that could not have been reported before. On the other hand, it would be difficult to verify these slides in the field. On this concern, the detected landslides were compared with a recently produced landslide susceptibility map for Trinidad (Baban et al. 2006). This has substantiated the possibility of the detected landslides to occur.

5.4 Conclusions

The chapter has provided a synopsis of the recent developments in acquiring geo-information using aerial and satellite-based remote sensing technologies and their utilization for investigating landslides. The high-resolution images, created by high-resolution satellite sensors (for example, 0.6–1.0 m) and ultra high-resolution airborne digital cameras (for example, 0.05–0.2 m), are becoming available and affordable. These data provide real opportunities for applications at time frames, resolutions and scales that were deemed impossible just a few years ago.

The Caribbean region can be characterized as mountainous, small islands with fast rates of development that can perpetuate rapid environmental degradation, but they have little or no up-to-date information for reliable and effective decision making. This is especially the case when dealing with landslide hazard management and mitigation. Therefore, the aforementioned technological developments are critical for the region as they provide opportunities for bridging the gaps in data and information needed for planning and management in terms of the time and space dynamics of the environment. More specifically, they provide effective means for surveying, inventorying, mapping and monitoring developments and the environment. Furthermore, they can be utilized to provide the necessary land parameters to run conventional landslides mathematical models as well as developing plausible scenarios to simulate environmental response to different natural events and development activities.

References

Alberta Environment. 2000. Geotechnical hazard assessment: South flank of Frank Slide. Hillcrest, Edmonton, Alberta.

Al-Tahir, R., and N. Thompson. 2007. The use of photogrammetry for landslide inventory in tropical environment. In press.

Al-Tahir, R., S.M.J. Baban and B. Ramlal. 2007. Utilizing emerging geo-imaging technologies for the management of tropical coastal environments. *West Indian Journal of Engineering* 29, no. 1:11–21.

Baban, S.M.J., F. Canisius, K. Sant and A. Chinchamme. 2006. Technical inputs to the Hillside Development Policy Technical Working Group. Ministry of Planning and Development, Government of the Republic of Trinidad and Tobago.

Baban, S.M.J., and K.J. Sant. 2005. Mapping landslide susceptibility for the Caribbean island of Tobago using GIS, multi-criteria evaluation techniques with a varied weighted approach. *Caribbean Journal of Earth Sciences* 38:11–20.

Ciciarelli, J. 1991. *A practical guide to aerial photography with an introduction to surveying*. New York: Van Nostrand Reinhold.

Corbley, K. 2000. Image processing and analysis: Empowering users with new tools. *Imaging Notes* 15, no. 3:18–20.

Cruden, D.M. 1991. Simple definitions of a landslide. *Bulletin of the International Association of Engineering Geology* 43:27–29.

Crystal, S. 2003. Trends in photogrammetry. *GIS Development* (March).

Dai, F., C. Lee and Y. Ngai. 2002. Landslide risk assessment and management: An overview. *Engineering Geology* 64:65–87.

Digital Globe. 2003. *QuickBird Imagery products and product guide.* Revision 4. Colorado: DigitalGlobe.

Ehlers, M. 2004. Remote sensing for GIS applications: New sensors and analysis methods. In *Remote sensing for environmental monitoring, GIS applications, and geology,* vol. 3, ed. M. Ehlers, H. Kaufmann and U. Michel, 1–13. *Proceedings of SPIE* 5239.

Hervás, J., J. Barredo, P. Rosin, A. Pasuto, F. Mantovani and S. Silvano. 2003. Monitoring landslides from optical remotely sensed imagery: The case history of Tessina landslide, Italy. *Geomorphology* 54:63–75.

Hinz, A., C. Dörstel and H. Heier. 2001. DMC: The digital sensor technology of Z/I-Imaging. In *Photogrammetric Week 01*, ed. D. Fritsch and R. Spiller. Heidelberg: Wichmann Verlag.

Karsli, F., A. Yalcin, M. Atasoy, O. Demir, S. Reis and E. Ayhan. 2004.

Landslide assessment by using digital photogrammetric techniques. *Proceedings of the Twentieth ISPRS Congress, Commission 7*, 736–39. Turkey.

Keating, T., P. Garland and C. Dörstel. 2003. Photogrammetry goes digital. *GIS Development* (March).

Leica. 2002. *ADS40 Airborne Digital Sensor*. Atlanta: Leica Geosystems, GIS and Mapping.

Li, R. 1998. Potential of high-resolution satellite imagery for national mapping products. *Photogrammetric Engineering and Remote Sensing* 64, no. 12:1165–69.

Mantovani, F., R. Soeters and C. van Westen. 1996. Remote sensing techniques for landslide studies and hazard zonation in Europe. *Geomorphology* 15:213–25.

MDA. 2006. *Radarsat-2: A new era in synthetic aperture radar*. Richmond, B.C.: MacDonald Dettwiler and Associates Ltd, Geospatial Services.

Metternicht, G., L. Hurni and R. Gogu. 2005. Remote sensing of landslides: An analysis of the potential contribution to geo-spatial systems for hazard assessment in mountainous environments. *Remote Sensing of Environment* 98:284–303.

Morena, L., K. James and J. Beck. 2004. An introduction to the RADARSAT-2 mission. *Canadian Journal of Remote Sensing* 30, no. 3:221–34.

Murphy, W., and R. Inkpen. 1996. Identifying landslide activity using airborne remote sensing data. *GSA Abstracts with Programs* A-408:28–31.

Orbimage. 2003. *OrbView-3 satellite and ground systems specifications*. http://www.orbimage.com/corp/orbimage_system/ov3/.

Rott, H., B. Scheuchl, A. Siegel and B. Grasemann. 1999. Monitoring very slow slope movements by means of SAR Interferometry: A case study from a mass waste above a reservoir in the Ötztal Alps, Austria. *Geophysical Research Letters* 26, no. 11:1629–32.

Schuster, R. 1996. Socioeconomic significance of landslides. In *Landslides: Investigation and mitigation*, ed. A.K. Turner and R.L. Schuster, 12–35. Transportation Research Board Special Report, no. 247. Washington, DC: National Academies Press.

Singhroy, V. 2002. Landslide hazards. In *The use of earth observing satellites for hazard support: Assessments and scenarios*. Final report of the CEOS Disaster Management Support Group, 97–114. Washington, DC: National Oceanic and Atmospheric Administration.

Singhroy, V., and K. Mattar. 2000. SAR image techniques for mapping areas of landslides. *Proceedings of the Nineteenth ISPRS Congress*, 1395–402. Amsterdam.

Singhroy, V., K. Mattar and L. Gray. 1998. Landslide characterization in Canada using interferometric SAR and combined SAR and TM images. *Advances in Space Research* 2, no. 3:465–76.

Singhroy, V., and K. Molch. 2004. Characterizing and monitoring rockslides from SAR techniques. *Advances in Space Research* 33, no. 3:290–95.

Soeters, R., and C.J. van Westen. 1996. Slope instability recognition, analysis and zonation. In *Landslides: Investigation and mitigation*, ed. A.K. Turner and R.L. Schuster, 129–77. Transportation Research Board Special Report, no. 247. Washington, DC: National Academies Press.

Space Imaging. 2003. *IKONOS Imagery products and product guide.* Version 1.3. Colorado: Space Imaging.

Spiker, E., and P. Gori. 2003. National landslide hazards mitigation strategy: A framework for loss reduction. Circular no. 1244. US Geological Survey, US Department of the Interior.

Tao, C. 2002. Digital photogrammetry: The future of spatial data collection. *GeoWorld*, no. 5:30–36.

van Westen, C. 2004. Geo-information tools for landslide risk assessment: An overview of recent developments. *Proceedings of the Ninth International Symposium on Landslides*, 39–56. London: Balkema.

Z/I Imaging. 2005. *Digital Mapping Camera System.* Huntsville, AL: Z/I Imaging Corporation.

SECTION 2

Floods

CHAPTER 6

Using GIS for Flood Management and Mitigation in Trinidad and Tobago

BHESHEM RAMLAL

Abstract

Trinidad and Tobago is plagued by a recurrent flooding problem. The higher levels of rainfall in the wet season often lead to extensive flooding in most parts of the country; this in turn leads to significant loss of livestock, a fall in agricultural produce, and damage to homes and businesses in many watersheds in Trinidad and Tobago. The disruptions caused by flooding in the last decade have affected productivity levels in most sectors of the national economy including agriculture, tourism, transportation, manufacturing, and energy exploration and production. Clearly there is a need for developing flood mitigation and management strategies to manage flooding in the areas most affected. This chapter reports on the findings of a study that utilizes a geographic information system to analyse the major causes of flooding in the Caparo River Basin, Trinidad, to map the extent of the flooding, to estimate soil loss due to erosion and to estimate sediment loading in the Caparo River Basin. The results of the study confirm that flooding is caused by several factors including clear cutting of vegetative cover, especially in areas of steep slopes, narrow waterways, poor agricultural practice and uncontrolled development in flood plains. The results of this work may be applied to other watersheds in Trinidad and Tobago to develop strategies for flood mitigation and management of these areas.

6.1 Introduction

Flooding is a major problem in many countries including Trinidad and Tobago, Bangladesh, India, Guyana, Germany, and the United States. In the last few years, Trinidad and Tobago experienced several severe flooding events that have led to significant damage to livestock, agricultural produce, homes and businesses as well as disruptions in several sectors including the oil and gas exploration and production sectors. In the last 20 years, floods caused the death of many persons, destroyed agricultural produce worth over US$20 million, and damaged over 1,000 homes and businesses (H. Wall, GIS officer, Central Statistical Office, Trinidad and Tobago, personal communication, 2006; Bryce 1999).

In Trinidad and Tobago, land-use practices such as slash-and-burn agriculture, quarrying, illegal logging, forest fires and illegal settlements have led to soils that are exposed to erosion, especially during periods of heavy rainfall and subsequent runoff. Consequently, heavy sedimentation occurs in the river channels causing these channels to be reduced in size. In addition, the absence of vegetative cover produces much shorter lag times between rainfall and the water reaching the waterways, causing the already reduced channels to overflow and leading to massive floods (Dion 2002). There is an urgent need to introduce flood mitigation measures to ensure that vulnerable areas are protected. This may be achieved using a three-step approach. The first step is to identify the nature and extent of vulnerability of the areas under consideration. Next, determine the most appropriate mitigation measures that should be used to address the problem. Finally, these measures must be implemented and maintained.

A promising strategy available for identifying vulnerable areas is to use spatial analysis tools available in geographic information systems (GIS). GIS analysis may be developed to examine spatial and temporal patterns and find associations between various geographical factors (Mitchell 1999). Since flooding is a spatial phenomenon, GIS will allow the user to handle, manage and analyse the spatial data sets to identify the effects of various factors and to forecast possible consequences (Melesse and Shih 2000). In terms of the impacts of land use and land cover on flooding, GIS may be used to detect change as well as identify trends, both visually and statistically, between land-use changes and flooded areas (Mamat and Mansor 1999).

6.2 The Study Area

Trinidad and Tobago are the two southernmost islands in the Caribbean chain. The country is approximately 5,000 km^2 in area with a population of 1.3 million people. Figure 6.1 shows the country in the Caribbean context as well as the study area that was used to demonstrate the applicability of GIS to flood mapping and analysis.

The island of Trinidad comprises of three major ranges of hills – the Northern, Central and Southern Ranges, and two plains – the Caroni

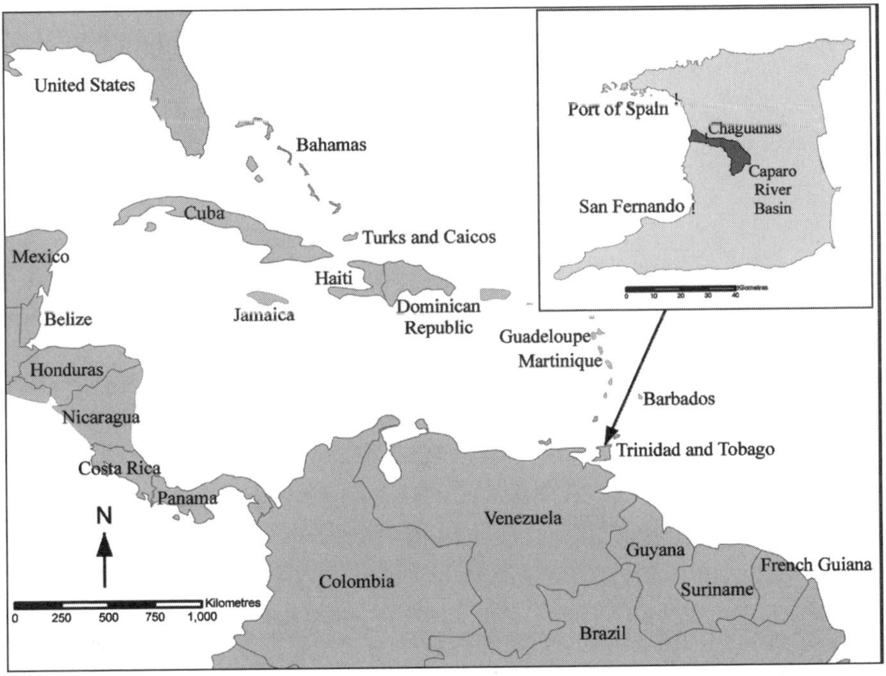

Figure 6.1 Location of study in the national and regional context.

and Oropuche Plains. The hills on the northern part of the island are much higher and steeper than the other two ranges. Figure 6.2 shows an elevation map of Trinidad. The ranges and plains may be easily identified. Major land use includes forested areas, which are plagued by illegal logging activities; commercial activities; residential areas, including areas of illegal settlements; agricultural activities; including slash-and-burn agriculture; swamps; and industrial activities, including illegal quarrying (Figure 6.3).

The major population is centred on the southern corridor of the Northern Range, running from the west coast of the island to the central parts. In the last few years, settlements have headed further north and higher into the watersheds of the Northern Range. As such activities on the island increase, the natural protective cover of the land is reduced, causing more flooding and erosion especially in the steep watersheds. This has led to extensive damage and loss of valuable raw materials and resources. Most of the areas flooded are used for residen-

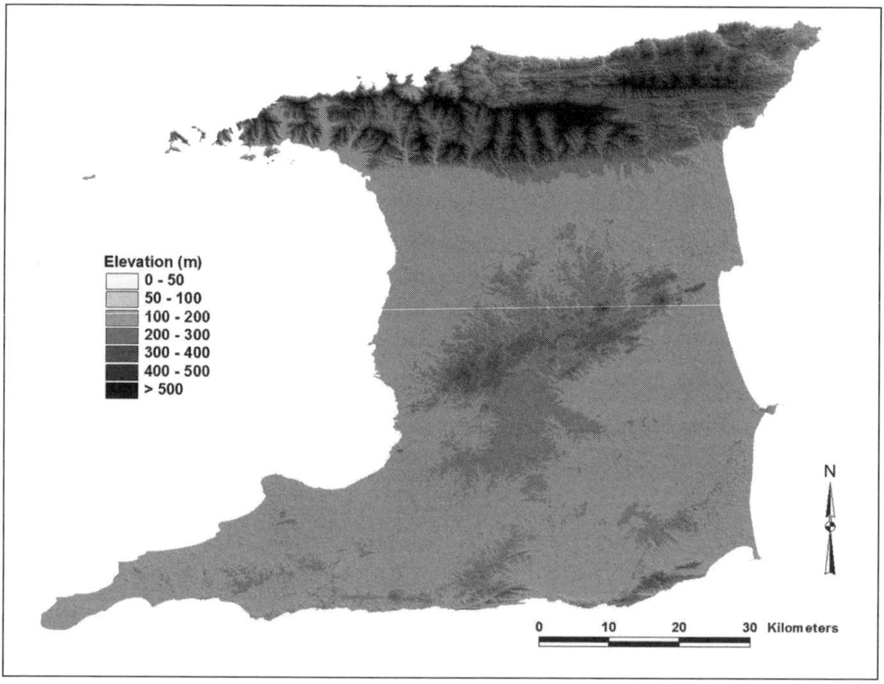

Figure 6.2 Elevation map showing ranges and plains.

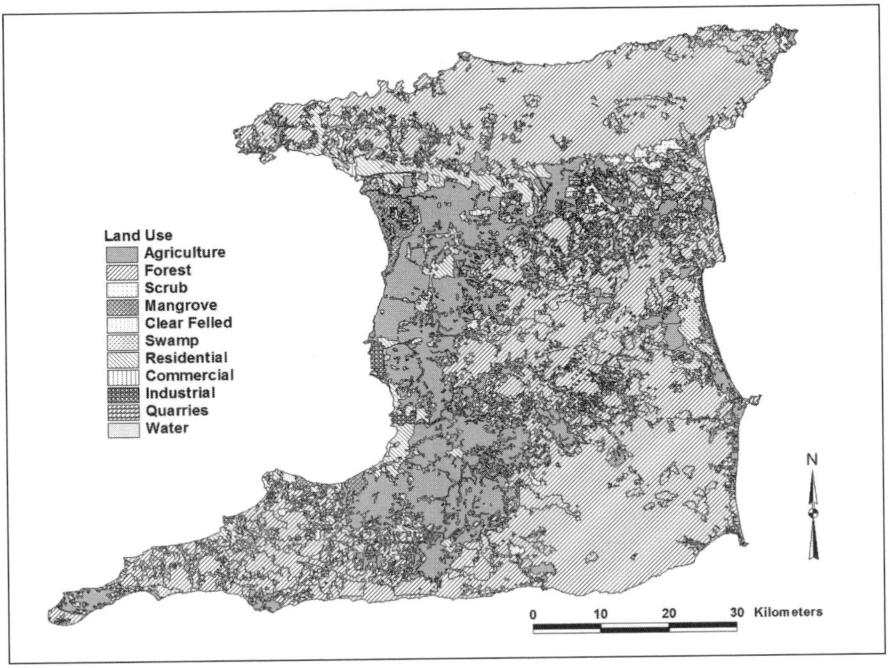

Figure 6.3 Land-use map of Trinidad.

tial and agricultural purposes. The problem is therefore quite significant for the entire country. To avoid future loss, it is necessary to identify the causes of the problem and attempt to address them urgently. Flooding has become such a problem in Trinidad in the last decade that the government recently commissioned projects to address flooding in more than ten major watersheds (personal communication, Public Relations Office, National Infrastructure Development Company, 2006). These projects follow a study conducted by a team including the author (Stere et al. 1999) to address flooding in the Caparo River Basin in Central Trinidad. See Figure 6.1 for the location of the Caparo River Basin in the national context.

The Caparo River Basin study was conducted on behalf of the Ministry of Works and Transport, Drainage Division, to develop flood mitigation and management strategies. The major results of this study are presented in this chapter. This chapter provides a discussion of the methodology used in the GIS component of the study to generate estimates of soil erosion and sediment yields and the identification of areas needing conservation and the type of conservation appropriate for these areas. Major findings are discussed. Conclusions and recommendations for addressing flooding at the national level are presented.

6.3 Data Collection Strategy for the Caparo River Basin Study

A thorough analysis of the existing conditions in the Caparo River basin was conducted to provide a basis for developing the solutions required. This involved the collection of historical data, including flooding events, surveys of land use and vegetative cover, rainfall, soil types, geology, elevation, population distribution, roads, river network system, including river profiles and cross-sections, bridges and dams, and delineation of the watersheds and sub-catchments for the upper watershed.

Figure 6.4 shows the elevation, land use/land cover, population distribution and soils maps as examples of the base maps used in the study. The dominant land uses were forests (35%) and agricultural activities (45%). The total population of the watershed is approximately 37,000 persons. The watershed elevation ranges from sea level to 270 m with more than 78% of the area lower than 100 m above sea level and 94% of the area with slopes less than 30°. The predominant soil types are Talparo – clay (20%), Brasso – clay (15%), Ecclesville – clay shale (11%) and Cunupia – fine sandy clay (10%). The average annual rainfall for the upper Caparo watershed area is approximately 2,000–2,200 mm, which decreases to about 1,600 mm in the coastal area. These data were summarized from the spatial data sets compiled for this project and were used in analysing the river system and its catchment.

To provide effective mitigation and management measures and to avoid significant erosion and flooding, it is important to analyse the soil erosion, sediment transport and river morphological processes of the Caparo River and its catchment (Dion 2002). A morphological assessment of the Caparo River was undertaken, and a sediment transport predictor was derived. GIS was also used to model flooding levels in the river basin for different events for 1-, 5-, 10- and 50-year return periods. Based on the findings of these analyses, recommendations on the

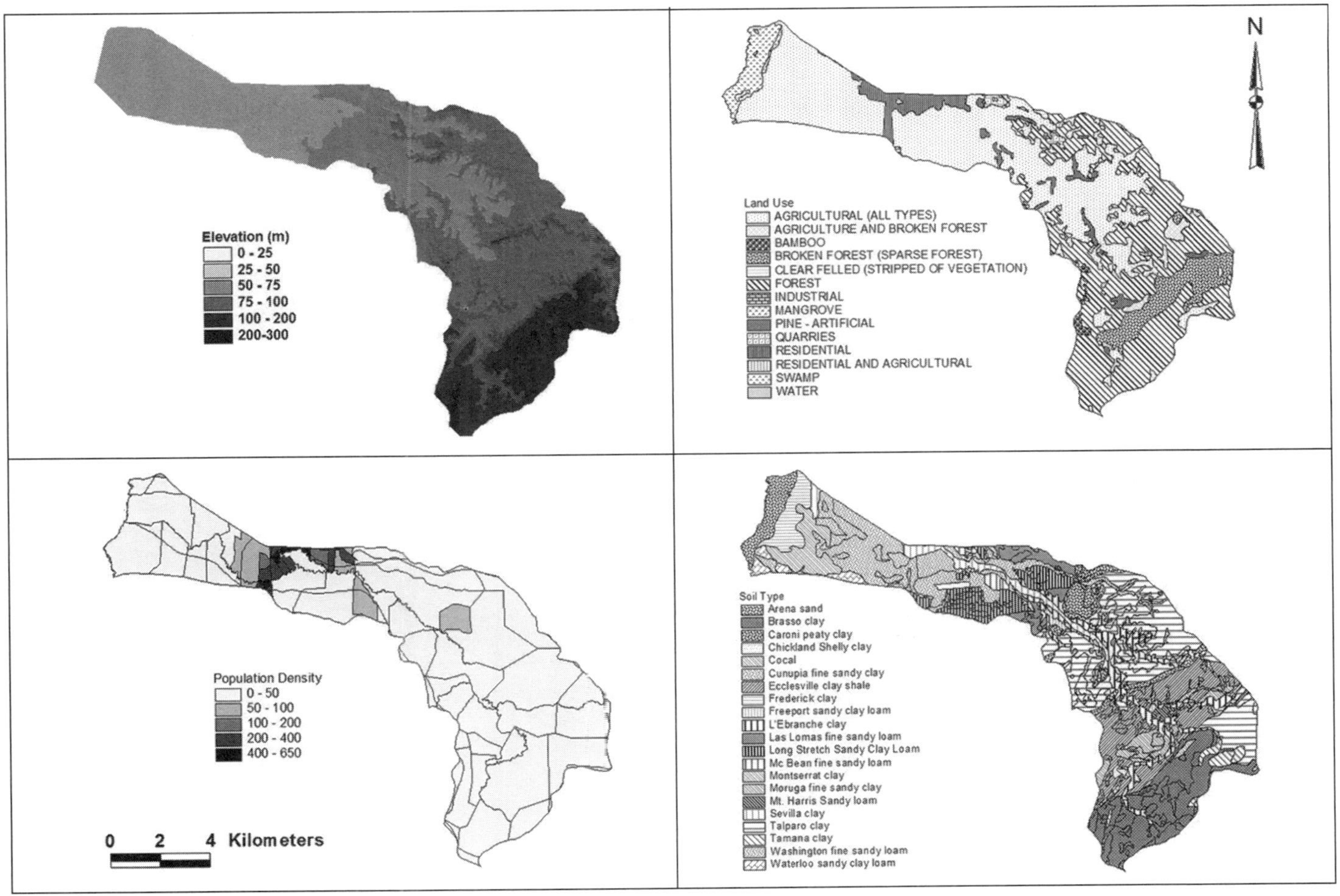

Figure 6.4 Spatial characteristics of the Caparo River Basin.

best solutions to mitigate flooding and manage the watershed are presented (Stere and Hesterman 1999).

6.3.1 The River System

The analysis of the specific characteristics of the river is based on the assumption that the Caparo river system is typically an input-output system (Figure 6.5). This implies that the output of the system is completely determined by the inputs, which are the basin characteristics and human interference. It should be noted from the system diagram that there are independent and dependent variables.

The independent variables occur at two levels: the catchment level and the level of the reach (Morisawa 1985). At the catchment level, the independent inputs are the climate and geology of the basin. The climate determines the average precipitation and temperature. The geological history is responsible for the rocks that are present and subject to weathering that leads to soil erosion. The influence of geology and climate is complicated by the role of vegetation and the weathering

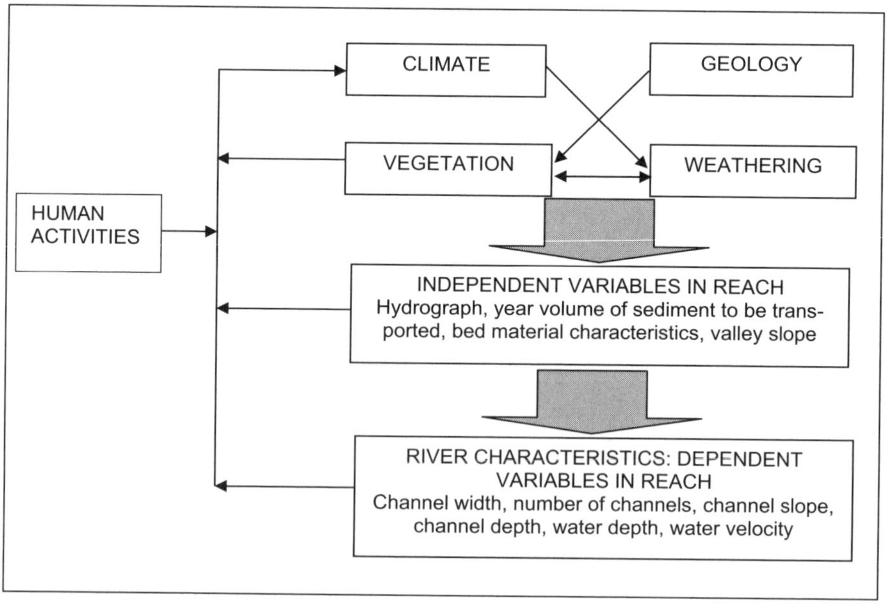

Figure 6.5 Variables defining river characteristics at different levels (after Stere and Hesterman 1999)

processes. More detailed discussions on the interaction between geology, climate, vegetation and weathering may be found in Morisawa (1985) and Richards (1987). At the river reach level, the independent inputs are the discharge hydrograph, the volume of sediment that has to be transported through the river reach on an annual basis, the characteristics of the sediments, whether coarse or fine sediments, and the slope of the valley. It may be noted that each of these variables will be different for each river reach. The dependent variables for a river reach are the river characteristics. These may be broken into morphological and hydraulic characteristics. These include bed material characteristics, longitudinal slope of the river, channel width, number of channels, channel slope, channel depth, water depth and water velocity in the river (Figure 6.5).

6.4 Estimating Soil Loss and Sediment Yield

Soil erosion is defined here as the amount of soil loss from a given slope, usually predicted per unit area basis, and sediment yield is the amount of sediment that passes a given point on the watershed. Some of the sediment that leaves a certain slope is deposited; hence, sediment yield and soil erosion are not the same (Haan et al. 1993).

Soil loss can be estimated using the universal soil loss equation (USLE) (Wischmeier and Smith 1978) which lends itself to GIS analysis. Improvements to this have resulted in a modification known as the revised USLE or RUSLE (Renard et al. 1997). The RUSLE is often used to identify areas that have already suffered damage and those which may be susceptible to damage if not managed properly. In this case the RUSLE is more appropriate since it better estimates the average annual soil loss from runoff for specified cropping and management systems (Dion 2002). The Caparo River Basin falls into this system.

The USLE/RUSLE equation is as follows: $A = R*K*L*S*C*P$

Where

A = average soil loss per unit area

R = rainfall/runoff factor

K = soil erodibility factor

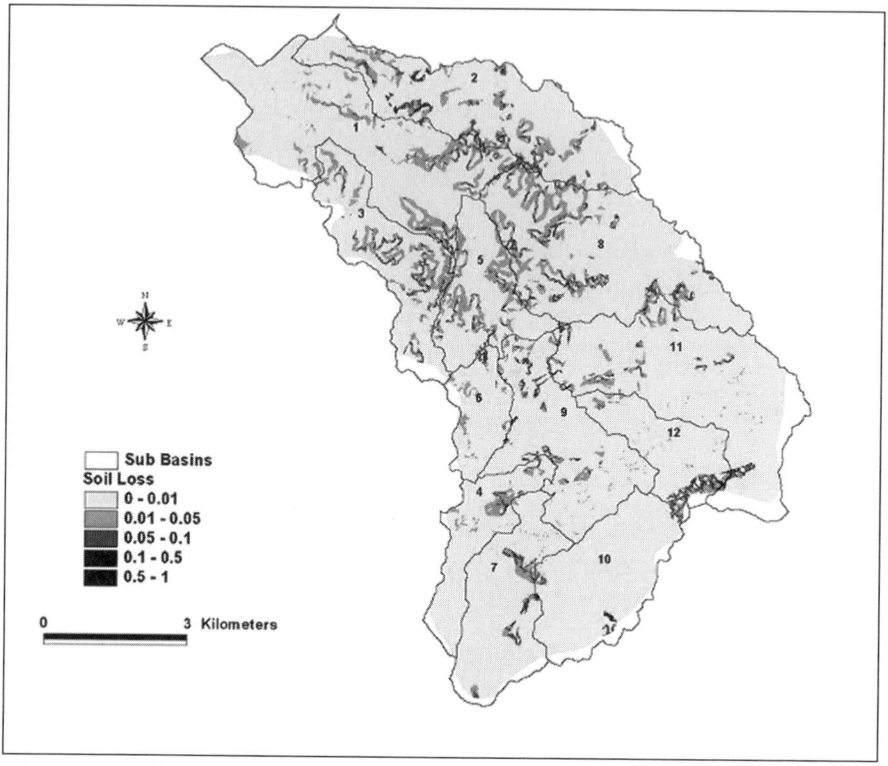

Figure 6.6 Soil loss estimation for the Upper Caparo River Basin Sub-catchments (cubic metres/square kilometres). The blank areas on the boundaries of the map represent data gaps.

L = slope length factor

S = slope steepness factor

C = cover and management factor

P = supporting conservation practice

For this study, the area was divided into the upper and lower watershed. The upper portion of the watershed was then broken into 12 sub-basins as shown in Figure 6.6. The lower basin was not analysed since the watercourse intervention was required at the lower edge of the upper basin. The results of the soil loss equation analysis for the sub-basins are presented in Figure 6.6. It may be noted that the highest level of soil loss occurs predominantly in the areas where the land use was agriculture. While steepest slope also contributed, this was less influential than land use. Similarly, the soil type does not contribute as significantly as the land use.

6.4.1 Sediment Delivery Ratio

Not all of the soil eroded in the catchment enters the river channel. Some of this soil is re-deposited in the catchment again. The sediment delivery ratio (D) indicates the difference. This was determined using a graph developed by Vanoni (1975) that relates D with the catchment size. The graph shows the average relationship together with the maximum and minimum values. The sediment delivery ratio (D) for each sub-catchment is presented in Table 6.1. R is the runoff factor.

6.5 Morphological Assessment of the Caparo River

A morphological assessment of the Caparo River allows an evaluation of the impact of improvement works on the river and an assessment of its morphological response. The proposed work will disturb the morphology of the river, causing it to adjust towards a more natural condition again.

The assessment required the collection of historical data and field data. Historical data are needed to assess changes that have occurred in the past. Field collection techniques were used to acquire topographic data as well as cross-sections of the river. In addition, bed samples were taken from 12 cross-sections of the middle and lower reaches of the river network and analysed to determine the bed material characteristics. Measurements of the cross-sections of the river were analysed. The average depth of the river is 4 m, and the width varied from about 10 m to 20 m or about 15 m on average. It was found that the bed material in the Caparo River changes downstream from fine sand to silt. In most of the cross-sections, clayey silt is observed. Overall, the river seems to be governed by depositional processes in the middle and lower reaches. This appears to coincide with the increased soil losses due to the changes in land use over the last century.

Away from the upstream hilly reaches, a river determines its own

Table 6.1 RUSLE Factors and Sediment Delivery Ratio for the 12 Sub-catchments of the Caparo River Basin

Catchment	1	2	3	4	5	6	7	8	9	10	11	12
Area	5.63	10.80	9.02	5.57	4.32	10.59	2.59	6.57	10.74	3.80	3.52	6.36
K-factor	0.042	0.036	0.038	0.040	0.038	0.040	0.042	0.041	0.041	0.042	0.040	0.042
S-factor	2.02	0.462	0.786	0.678	0.57	0.786	1.012	1.348	1.012	1.516	1.348	2.188
L-factor	2.83	1.82	2.13	2.00	1.88	2.13	2.31	2.67	2.32	2.62	2.45	2.86
C-factor	0.027	0.159	0.103	0.135	0.137	0.083	0.036	0.056	0.042	0.050	0.028	0.012
P-factor	1	1	1	1	1	1	1	1	1	1	1	1
Soil loss indicator A/R	.0065	.0049	.0066	.0073	.0056	.0056	.0035	.0082	.0040	.0083	.0037	.0031
A/R* area	0.036	0.053	0.059	0.041	0.024	0.060	0.009	0.054	0.043	0.030	0.013	0.019
Sediment delivery D	0.2	0.17	0.185	0.2	0.21	0.17	0.4	0.195	0.17	0.22	0.22	0.195
D* A/R* area	0.0072	0.009	0.011	0.008	0.005	0.01	0.003	0.01	0.007	0.007	0.002	0.004
Contribution to D (%)	8.4	10.5	12.9	9.6	5.9	11.9	4.3	12.3	8.6	7.7	3.4	4.4

slope, planform and cross-sectional characteristics (Dion 2002). Data was also collected to determine these characteristics of the river. The average slope of the river varies from 3.5 m/km in the upstream reaches to 2 m/km in the downstream reaches. In the upstream reaches, the planform of the river is dictated by the surrounding hills. Several tight meanders have formed. From midstream onwards, the river exhibits the character of a river meandering in a wider valley. In most reaches, the river is a well-defined single channel. Only a small part of the downstream has several channels most likely caused by diversions for the construction of railway lines and roads. Often the width of the channel can be linked to the discharge the river conveys when the flood plains start to be inundated. However, the vegetation along and on the banks of the river may significantly influence the discharge, the sediment transportation width and the shear stresses on the riverbed and hence on sediment transport. The vegetation cover was analysed to determine the extent of influence. It was found that vegetation has a pronounced effect (Stere et al. 1999).

6.6 Developing an Integrated Approach to Flood Management in the Caparo River Basin

The above information was used to identify the sub-catchments where erosion was the most severe and the factors responsible, to identify the waterways that are most affected by the sediment loading because of the erosion, and to determine the likely outcome of mitigation measures that are proposed. The results obtained were used to develop plans for flood management in the Caparo River Basin. These are presented below.

The Watershed Management Plan

The GIS was used to execute queries and analyses to identify the nature and the extent of the problems in the watershed. The identified issues included the need for reforestation of steep slopes; the introduction of conservation measures for areas used for agricultural purposes; and the need to control further development in the lower and middle reaches, especially in the flood-prone areas. The areas identified for conserva-

Table 6.2 Criteria Used in Identifying Areas Needing Conservation

Slope	Soil Type	Area (km2)	Conservation Measure
Greater than 20°	All types	0.4	0.01km^2 not under forest cover – all areas need to be covered with forest
Between 5° and 20°	All types	21.7	Measures depend on soil type and land use
Between 5° and 20°	68/L, 71, 182/L, 241, 261, 482/L, 575, 674/L	8.2	Need to be under vegetative cover; either forest or agriculture with soil conservation measures
Between 5° and 20° and under agriculture	68/L, 71, 182/L, 241, 261, 482/L, 575, 674/L	0.6	Soil conservation measures needed

tion were based on the slope and soil types. Table 6.2 shows the different criteria for areas identified in the upper basin and the measures recommended for conservation. Conservation measures included the use of earth bounds, contour drains, contour planting, vegetable barriers, diversion channels, graded drains and strip cropping. In the middle and lower parts of the river basin, the major issue was that of expansion of development into the flood plains. Further development should be tightly controlled by the relevant agencies. Institutional arrangements need to be developed to ensure that agencies work together in managing watersheds.

Flood Control Works

The flood control works proposed include the upgrading of channels to increase conveyance capacities of channels and the construction of a retention pond to hold back the high discharges that enter the flatter valley floor to release at a rate that is suitable for the upgraded downstream channel capacity. In addition, the bypass channel around the town of Chaguanas should be reopened to divert the greater part of the flood-flows away from the town. A trash rack should be placed upstream of the town so that trash is collected before the flow enters the urban area. Finally, lands should be allocated to store and convey floodwaters on rare occasions when floods are exceedingly high.

The Land Acquisition Plan

Several tracts of land would be required to install the diverse infrastructure needed to mitigate flooding in the Caparo River Basin. A corridor along the general course of the Caparo River is needed for engineering works. In addition, an area close to the town of Mamoral is needed to construct a detention basin. The extent of this area was identified using a flood hydrograph of 1:50 years return period. This area is inhabited, and most of the lands are privately owned. A land acquisition strategy along with relocation plans for both existing occupants and a cemetery were developed to secure the lands needed.

The Environmental Management Plan

Mitigation measures are proposed to minimize the negative impacts and maximize the beneficial impacts of all works to be introduced. The following mitigation measures were recommended: (1) increase in vegetative cover through reforestation; (2) more educated and effective use of pesticides to mitigate negative impacts; and (3) undertake dredging and construction work in the dry season to avoid siltation and sediment transport due to runoff.

The Operation and Maintenance Plan

Since major investments are being made, it is necessary to ensure that there is continued effectiveness in the infrastructure being introduced to the project area. The following are recommended: (1) establish a dam safety unit with personnel with appropriate training and resources to conduct all aspects of dam safety and maintenance monitoring; and (2) provide adequate resources and develop strategies to ensure that detention basins, river alignments, embankments, river crossings, diversion structures and flap gates are all well maintained.

6.7 Discussion and Conclusions

Flooding in the Caparo River Basin is caused by many factors, including exposed soils, quarrying activities, slash-and-burn agriculture and silted drains. For this study, GIS was successfully used in determining the extent of the problem, including the identification of areas that needed conservation, the extent of soil loss and flooding at different sections of the watershed. The analysis provided by GIS made it possible to expedite the development of mitigation strategies that are most likely to address the anticipated changes in the river basin. Flooding in the Caparo River Basin may be mitigated if the recommendations provided are implemented in their entirety since the proposed strategies comprise of many interdependent components. Piecemeal implementation will not provide the appropriate results. It is recommended that the methodology developed for this study be applied for all the watersheds in Trinidad, especially since most of them are similar in extent, development levels, and physical and other characteristics.

References

Bryce, R. 1999. Report of the national coordinator of the Caribbean Land and Water Resources Network CLAWRENET and hydrologist at the Ministry of Agriculture, Land and Marine Resources MALMR, Trinidad and Tobago. http://www.procicaribe.org/networks/clawrenet/reports/z_tt/tt.htm.

Dion, T.R. 2002. *Land development for civil engineers*. 2nd edition. New York: John Wiley and Sons.

Haan, C.T., B.J. Barfield and J.C. Hayes. 1993. *Design hydrology and sedimentology for small catchments*. San Diego: Academic Press.

Mamat R., and S.B. Mansor. 1999. Remote sensing and GIS for flood prediction. *Proceedings of the Asian Association of Remote Sensing*. http://www.gisdevelopment.net/acrs/1999.

Melesse, A.M., and S.F. Shih. 2000. Geomorphic GIS database for runoff coefficient determination. *Proceedings of the Second International Conference on Geospatial Information in Agriculture and Forestry*, vol. 1, 505–12. Lake Buena Vista, Fla.

Mitchell, A. 1999. *The ESRI guide to GIS analysis*. Vol. 1. Redlands, CA: ESRI Press.

Morisawa, M. 1985. *Rivers*. New York: Longman.

Renard, G.G., G.R. Foster, D.K. McCool and D.C. Yoder. 1997. *Predicting soil erosion by water: A guide for conservation planning with the revised universal loss equation RUSLE*. Agriculture Handbook no. 703. Washington, DC: US Deptartment of Agriculture, Agriculture Research Service.

Richards, K.S. 1987. *River channels: Environment and processes*. Institute for British Geographers, special publication no. 18. Oxford: Blackwell.

Stere, C., and Hesterman, E. 1999. Caparo River Basin Flood Mitigation and Water Resources Management Project: Feasibility report and implementation plans. Report to the Government of Trinidad and Tobago.

Stere, C., E. Hesterman and S.J. Visser. 1999. Caparo River Basin Flood Mitigation and Water Resources Management Project. Vol. 4. Report to the Government of Trinidad and Tobago.

Vanoni, V.A. 1975. *Sedimentation engineering*. New York: ASCE.

Wischmeier, W.H., and D.D. Smith. 1978. *Predicting rainfall-erosion losses: A guide for conservation planning*. Agricultural Handbook no. 537. Washington, DC: US Department of Agriculture.

CHAPTER 7

Using GIS for Flood Risk Assessment and Flood Sensitivity Maps for a Watershed in Trinidad and Tobago

SERWAN M.J. BABAN and RONNIE KANTASINGH

Abstract

In Trinidad and Tobago, flooding is a major perennial problem causing injury to persons, damage to infrastructure, economic losses and general destruction. The occurrence of flooding is not a one-off event, because major floods occur yearly. In 2002, there was a major flooding event in Valsayn, another in 2003, in the Maraval/Woodbrook area, and yearly within the Caparo River Basin. To effectively deal with this problem, there must be an understanding of the intricate relationships existing between the ecosystem and socioeconomic activities in river basins. Moreover, because the physical, climatic and environmental factors are unique to the region, existing flood risk plans and flood forecast models do not seem to be appropriate to use.

This chapter examined the factors that may have played a part in flooding in the St Joseph Watershed on 5 November 2002, including rainfall, deforestation, development/housing and squatting. As a result, a flood risk assessment map for the St Joseph Watershed was developed. Additionally, three sensitivity analysis maps were created to determine the influence of identified flood risk factors. The development of the flood risk assessment map and the flood sensitivity maps using geographic information systems (GIS) is aimed at providing authorities with tools for flood management, river network upgrading, land-use planning and watershed management.

7.1 Introduction

Of all the natural hazards that affect the spatial organization of mankind, none is as universally critical as flooding (Allan and Bradshaw 1981). A flood can be defined as "any relatively high stream flow overtopping the natural or artificial banks in any reach of a stream" (Leopold and Maddock 1954, 278), while the flood plain can be defined as "a strip of relatively smooth land bordering a stream, built of sediment carried by the stream and dropped in the slack water beyond the influence of the swiftest current" (Langbein and Iseri 1995). The main causes of flooding include the following:

Climatic effects: These include the combined effects of weather systems (cold or warm fronts, ITCZ), wind speed, humidity, pressure and so on.

Topographic effects: The presence of highlands can cause orographic or relief rainfall. In a study done in Taiwan, it was found that flooding is prevalent during the typhoon season because of intense and prolonged rainfall, coupled with orographic influences creating torrential runoff (Chan et al. 2003).

Human induced: Slash-and-burn methods used to clear vegetation leads to both the removed debris as well as eroded topsoil being washed into watercourses. These results in a reduced carrying capacity of the watercourse as well as possible blockage of out-flow points of watersheds. Additionally indiscriminate dumping and discarded litter also exaggerates the above problem. Figure 7.1 highlights the main causes of flooding and show the human influence on the chain of processes.

Regardless of the factors causing floods, their effects always include destruction, possible loss of life, and financial and emotional suffering. The general effects of floods and flooding are as follows:

- Spread of diseases (water-borne infections, vector borne diseases, rodent-borne diseases) and possible loss of life (drowning);

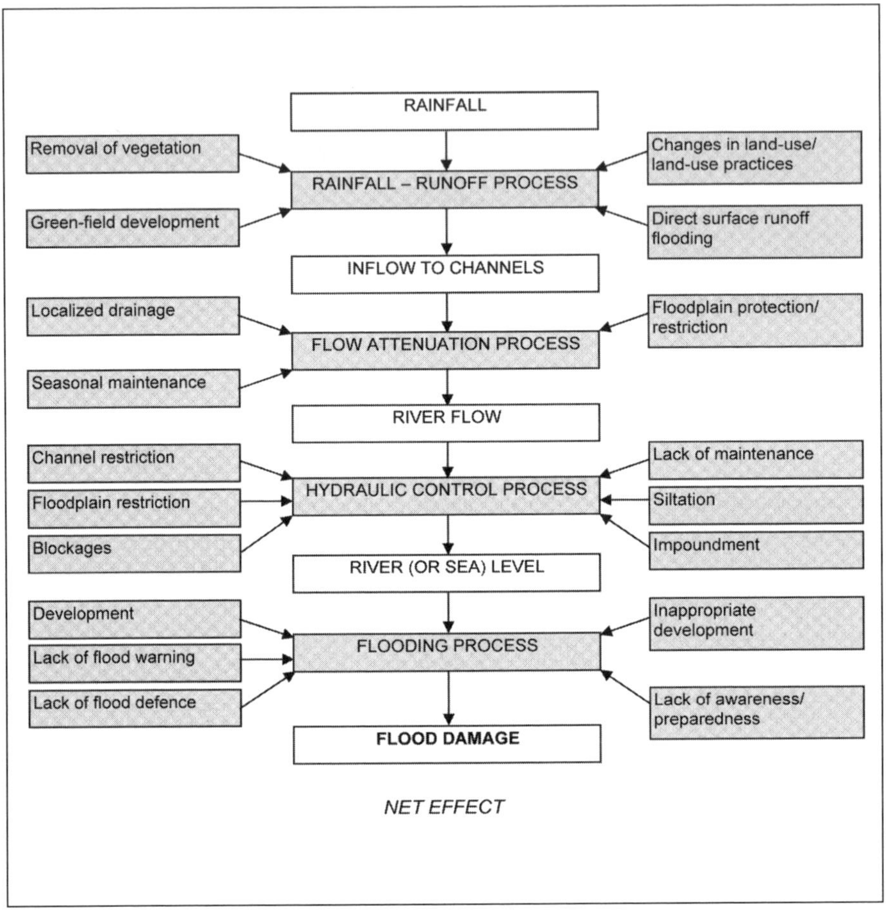

Figure 7.1 Causes of flooding – human impacts on the chain of processes (Adamson 2003).

- Economic losses for homeowners, businesses, farmers and government;
- Destruction may vary from flooded homes (damaged or destroyed furniture, clothes, appliances, food stocks), to flooded businesses (both private and state owned), flooded schools, churches and so on. Damage also occurs to roads, utility lines/poles, rivers and other physical structures, causing the delay of public services and the disruption of daily life;

- Agricultural (soil loss) and livestock losses, floods also wreak havoc on farmers (agricultural and livestock) since agricultural fields may be flooded, and entire crops lost. Additionally, flood waters tend to cause erosion, and nutrient rich topsoil may be washed away. They also cause loss and death of livestock since, on many occasions, the animals are either tied or in pens and drown because they cannot escape the rising flood waters;
- Physical and emotional stress to residents, business owners and citizens who suffer losses due to flooding; and
- Torrential downpours also result in floodwaters suddenly, and without warning, sweeping down hillsides and into rivers. This rush of water and debris causes secondary hazards like landslides and mudflows that cause destruction and possible deaths. Such was the case in Tobago, in November 2004, when two lives were lost due to floods and landslides. Therefore, flooding within tropical islands like Trinidad and Tobago does not only mean that areas may be submerged in contaminated waters.

As the population of a country grows, there is an additional strain on all the resources available. One of the major problems facing developing countries is the inability of governments to provide adequate housing and job opportunities for households to sustain themselves. In Trinidad and Tobago there has been a 0.6% population growth rate and a 1.9% annual labour force growth between 1996 and 2002 (World Bank 2003), while the gross domestic product has dropped from 2.2 in 1982 to 1.6 in 2001. As a result, many persons have resorted to squatting on unoccupied state owned lands, both for housing and farming. The Land Settlement Agency (LSA) of Trinidad and Tobago now estimates the number of squatter households to be about 25,000 (LSA 2001). These squatter developments rely on slash-and-burn techniques to clear land and, according to the Food and Agriculture Organization of the United Nations, deforestation in Trinidad and Tobago accounted for the loss of approximately 2,200 ha of natural forests, or 0.8% per year, during the period 1990–2000. An internal forestry department report in 2001 estimated that in 1994, approximately 11,593 ha or 8% of forest cover had been removed illegally (Forestry Report 2001). This indiscriminate removal of the earth's natural cover leads to increased surface runoff and soil loss, silted watercourses, and overall degraded watersheds and catchments. Trees protect the soil against the direct impact of rainfall. Their leaves not only shield the earth, but also help water drain to the ground in a controlled manner, and the roots help bind the soil to prevent erosion. This process, where the precipitation is temporarily stored by the vegetation, is called interception (Ward and Robinson 2000). In a 2000 report, the Food and Agriculture Organizat-ion indicated that poor land-use and land-management practices have entailed degradation of watersheds, forests, farms, aquifers, coastal zones, fisheries and coral reefs, all of which affect the Caribbean environment and its peoples.

The inability of state agencies to effectively deal with the problems of wrongdoers is shown in the example cited by the International Tropical Timber Council (2003): "With respect to squatting, the Forestry Department has only a weak mandate to expel squatters and the Sub-intendment of State Lands and Survey Department whose job it is to regulate squatting are under-staffed and have been reluctant to exercise their mandate." Presently, the existing legislation directly or indirectly affecting land management and land issues include the following (Ramkisoon 2000):

- The Environmental Management Act
- The Planning and Development of Land Bill
- The State Lands Regularization of Tenure Act
- The Agricultural Small Holdings Tenure Bill
- The National Parks and Other Protected Areas Bill
- The Municipal Corporations Act
- The Tourism and Industrial Development Company of T&T Ltd Vesting Order Act

This chapter seeks to show how GIS can be incorporated to identify and map the factors that may have played a part in flooding in the St Joseph Watershed on 5 November 2002, as well as the ways in which GIS can assist in the development of a flood risk assessment map for the St Joseph Watershed. Additionally, GIS capabilities will also be used to develop flood sensitivity maps to highlight the influence of individual flood risk factors.

7.2 GIS and Flooding

A GIS can be defined as "a powerful set of tools for collecting, storing, retrieving at will, transforming and displaying spatial data from the real world for a particular set of purposes" (Burrough and McDonnell 1998). A successful GIS consists of the combination of five in sync components (people, software, hardware, procedures and data) that seek to organize, automate, manage and deliver information through geographic presentation (Zeiler 1999). The rise of computers and, subsequently, new processing and analytical power have revolutionized the way scientists collect data as well as the way the information generated is used to identify trends/patterns and to develop management scenarios. This new level of thought, where the spatial attributes of objects are examined, has brought to the fore the field that is GIS.

GIS analyses are well suited for examining geographic patterns and detecting associations between features (Mitchell 1999). Since flooding is as a consequence of a number of factors, including soil type, vegetation cover and type, rainfall intensity, rainfall frequency and others, all of their influences have to be taken into account when analyses are conducted to determine what factors have what effect and the resulting consequences. GIS have the capabilities to store, retrieve and analyse different types of data for management of natural resources (Seth et al. 1999). GIS allows the spatial information to be displayed in integrative ways that are readily comprehensive and visual (Grover 1999). Furthermore, GIS and remote sensing have been used successfully in monitoring and managing land-use changes at the watershed scale through assessing the impact of land change and the resulting runoff problems (Sharma et al. 2001).

One of the strongest assets of a GIS is the ability to carry out temporal analysis. In hydrology, GIS can be used for flood prediction by storing data on previous floods, soil types, river channel size and so on. In addition, it can be used to create models of peak flow, discharge or runoff to determine what the consequences would be for a rainfall incident of a particular intensity and frequency. Within the Caribbean region, the lack of historical data as well as data sets not being available in digital format and the late emergence and acceptance of GIS technology have resulted in the slow development of major GIS projects. In a study carried out in West Bengal, a GIS-based flood-hazard map was prepared using limited data (Sanyal and Lu 2003). Sensitivity analysis is used to ascertain how a given model output depends upon the input parameters (Chan et al. 2003). By removing completely or varying the percentage influence of an input parameter, the output will vary. This functionality allows authorities to create multiple scenarios and models for future watershed/catchment management. As such, sensitivity analysis should be considered as a pre-requisite for statistical model building in any scientific discipline where modelling takes place (Satelli et al. 2000).

As a result of not being able to predict when and where the next flood will occur, the development of a flood risk assessment map and flood sensitivity maps are essential. Flooding cannot be completely avoided, but damages from severe flooding can be reduced if a flood of a particular magnitude and its impact can be predicted and an effective flood prevention scheme implemented (Aziz et al. 2002).

7.3 The Study Area

The St Joseph Watershed lays nestled in the Leeward side of the Northern Range in the island of Trinidad (Figure 7.2a). The watershed is located between 10°38’ and 10°44’ north latitude and 61°26’ and 61°23’ west longitude and is approximately 48.86 km^2. It runs from the hill slopes to the base of the Northern Range and onto the Caroni Plains. The highest point of the watershed runs along the 760 m contour line and the lowest areas of the study area are approximately 5 m to 10 m above sea level. The island experiences two relatively distinct seasonal types, tropical maritime and modified moist equatorial, resulting in two distinct seasons – a dry season (January to May) and a wet season (June to December), with the average minimum and maximum temperatures ranging between 22°C and 25°C and 29°C and 31°C respectively (EMA 2001). During the rainy season the island receives two-thirds of its 2,200 mm annual precipitation (EMA 1996).

The St Joseph Watershed lies within an area that has, for the most part of the island’s history, been covered with virgin, tropical forest. After slavery and indentureship, many persons left the sugarcane indus-

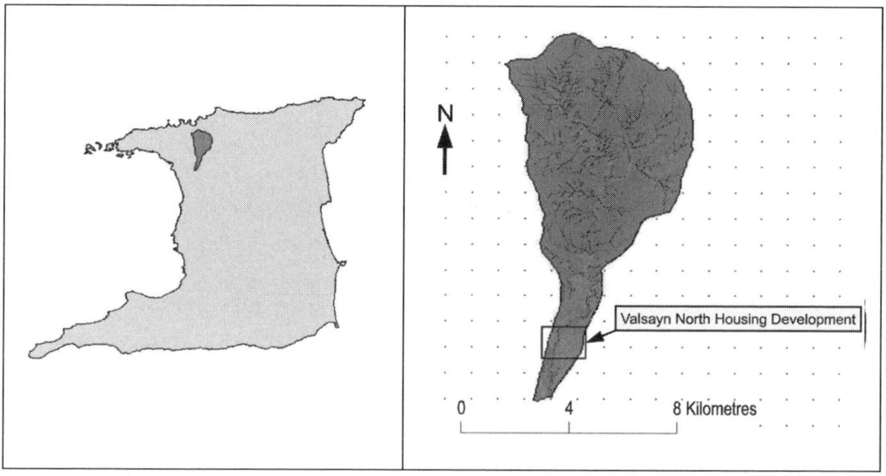

Figures 7.2a and 7.2b Trinidad and the St Joseph Watershed.

try and turned to the fertile hill-slopes that were "free" of charge and started planting cash crops. Later on, as the main city centre, Port of Spain became saturated with jobs and residential opportunities, both people and businesses moved to new areas. St Joseph was one of these places, and most of the housing was eventually located in and around the foothills, with the agricultural sector being found further up the mountain slopes. Field visits and visual inspections of aerial photos from 1984, 1998 and 2000 indicate clearly that the development within the St Joseph watershed has drastically increased, and with it, the number of natural forested and "natural green" areas has decreased. Additionally, the main road runs almost parallel to the St Joseph River. This allows rainfall water to quickly run off into the watercourse almost uninhibited and hence increases the risk of flooding (Kantasingh 2005).

The predominant soil group within the watershed is micaceous phyllites (78%), which is a high upland type soil with free internal drainage. The second largest soil type is mica phyllite sand (7%), which is a deep alluvial soil with free drainage (Baban and Kantasingh 2005) The St Joseph catchment (Figure 7.2b) is relatively thin and long with many tributaries flowing into the main St Joseph River, which eventually empties into the Caroni River. The flow of water down the hill slope and watershed is aided by the fact that the catchment is short and steep. In fact, the upper point of the watershed is approximately 1,000 m above sea level, and the lower regions are only about 4 m or 5 m above sea level (Kantasingh 2005). Additionally, the Caroni River also drains most rivers in the Northern Range of Trinidad, so that after any given rainfall event the level of water in the river will be somewhere near maximum. It should also be noted that the proximity of both these rivers to the Gulf of Paria means that at high tide, the level of water in the Caroni River, and to a lesser extent in the St Joseph River, will be higher than normal.

7.3.1 The Flood Event

The climatic conditions prior to 5 November 2002 indicate that the rainfall on 1 November to 4 November would have caused the mostly sandy clay loam soil to be thoroughly saturated, and then the 62 mm of rainfall within a 3–hour time period on 5 November would have simply run off the surface and onto watercourses towards the river (Figure 7.3). Additionally, the wind speed during the storm event on 5 November was found to be almost 0 m/s, which meant that the rainfall system would have been stagnant, and rainfall persistent, over the study area. As indicated previously, the river draining the watershed (the

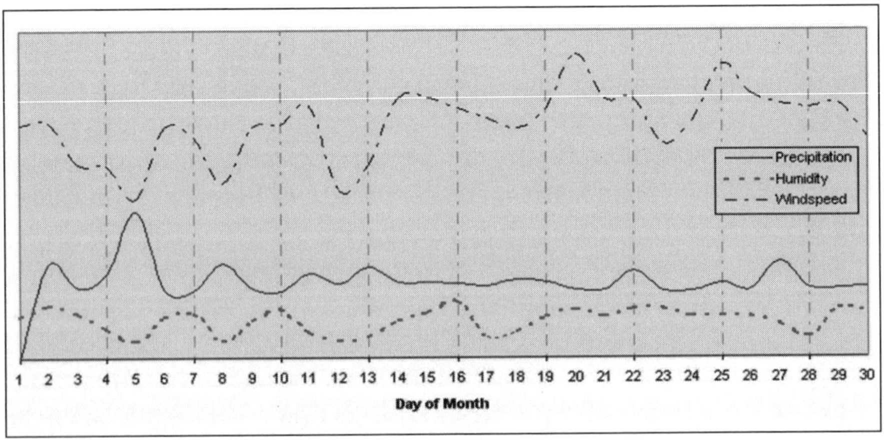

Figure 7.3 A comparison of precipitation, humidity and wind speed for November 2002.

St Joseph River) flows into the Caroni that would have been almost, if not completely full. Added to this was the fact that around 4:00 p.m. on 5 November, there was a high tide, and the Caroni River, which normally flows into the sea itself, would have had its level of water automatically raised. This would mean that the water from the St Joseph River was unable to find a point to empty and therefore overflowed its banks. This argument can be verified by using the rainfall event on 19 November 2003, when 105 mm was recorded for the same watershed – some 43 mm more (Kantasingh 2005; Baban and Kantasingh 2005). However, no flooding occurred since the rainfall prior was not as high, the soil not as saturated and there was a low tide at the time. Lastly, it was found, via field visits and from questionnaires, that the St Joseph River was not being properly or regularly cleaned. This litter, which included old appliances, tires, bags of rubbish, growing vegetation and other debris, not only acts as a health hazard but also reduces the carrying capacity of the river. It was found that such debris blocked the pathway under the bridges in Valsayn and contributed to the river overflowing its banks and causing flooding (Kantasingh 2005).

7.4 Methodology

7.4.1 Data Collection and Development

Table 7.1 shows the type of the data themes and the sources needed for their analysis, which exist in the form of hardcopy map sheets. These data sets were obtained and scanned using a flatbed scanner at 200 dpi, and the vectorization software R2V was used to digitize and geo-reference the 8-bit grayscale TIFF image. The output polygon shapefile was imported into the editing software DAK, and, using the "Clean" and "Built" commands, errors were identified and removed and topology added. The final polygons were saved as ArcView shapefiles. Within ArcView, and with the aid of 3D Analysis extension, the contour data layer, together with spotheights obtained from initial field surveys, were used to first create a TIN and then reclassified to form a slope grid theme.

Table 7.1 The Data Required, Data Sources and Data Themes Created

Theme	Sources and Initial Data Format	
River	Ministry of Works Drainage Division UWI CLEAR	1:25,000 Topo. map sheets
Soils	UWI CLEAR	1:25,000 Topo. map sheets
Land Use	Lands and Surveys Department UWI CLEAR	1:25,000 Topo. map sheets aerial photographs
Contour	UWI CLEAR Lands and Surveys Department	1:25,000 Topo. map sheets
Rainfall, Wind Speed, Temperature	Meteorological Office Piarco, UWI Field Station	Maps, tables and graphs
Tidal Data	Meteorological Office Piarco	Maps, tables and graphs
Spot heights	Field Surveys (GPS, Total Station)	Authors
Personal Experience	Residents of Valsayn	Questionnaire

7.4.2 Developing a Flood Risk Assessment Map

A basic way to create or identify spatial relationships is through the process of spatial overlay. Spatial overlay functions by joining and viewing together separate data sets that share all or part of the same area. The outcome from this process is a new data set that identifies the spatial relationships (Volusia 2001). The Model Builder extension within ArcView 3.2 was used to perform a weighted overlay analysis of the land use/land cover, soil type and slope grid themes to derive an output flood risk assessment map that identified areas within the watershed that may be susceptible to flooding (Figure 7.4). Unfortunately, data sets were only available for the three factors above. These factors carried the same flood-hazard weight (% influence), and rainfall was constant over the watershed (Table 7.2). The final map was verified

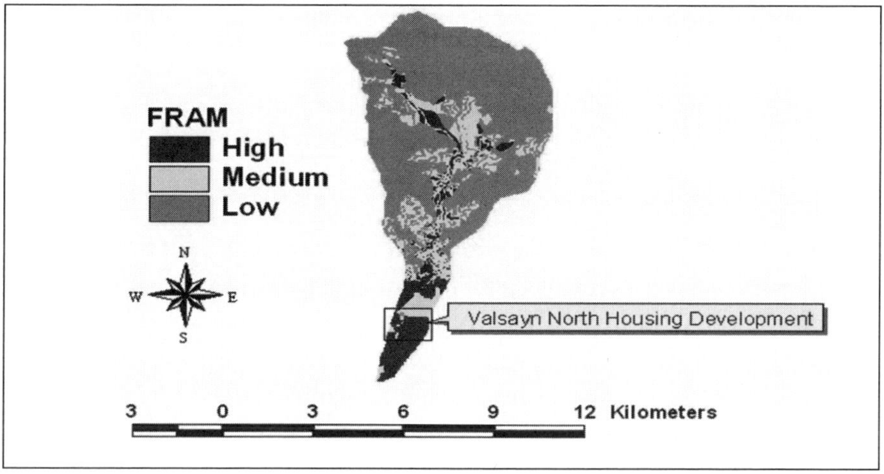

Figure 7.4 The flood risk assessment map (FRAM).

Table 7.2 Weighting Scheme Used for the Overlay Analysis

Input Theme	% Influence	Input Field	Input Label	Scale Value
Land-use Map	33	1	Built-up area	9
		2	Citrus	8
		3	Forest	1
		4	Other crop	6
		5	Savannah	7
		6	Scrub	5
		7	Vegetable garden	6
		8	Residential	9
		No data	No data	Restricted
Slope Map	33	1	0–17.266	9
		2	17.266–34.532	7
		3	34.532–51.798	5
		4	51.798–69.064	3
		5	69.064–86.330	1
		No data	No data	Restricted
Soil Map	34	1(fine sandy loam, sandy clay)	Low	9
		2 (sandy clay loam)	Moderate	5
		3 (clay loam, clay)	Very High	1
		No data	No data	Restricted

using data collected via questionnaires, as well as pictures taken by the research team the day after the flood event.

7.4.3 Developing Flood Sensitivity Maps for the Event on 5 November 2002

Three alternative sensitivity maps formulated to the final flood risk assessment map were developed. These three were made on the basis that the input data layers were not of equal significance to flood risk analysis, and the following percentages were used in the new weighting. The results of these three scenarios are shown in Tables 7.3 and 7.4 and Figure 7.5. The analysis of the results showed that the changes in flood risk were as shown above (Figure 7.5).

Table 7.3 Weighting Schema for the Sensitivity Maps

FRAM*	Scenario 1 (Map 1)	Scenario 2 (Map 2)	Scenario 3 (Map 3)
Land Use = 33.3%	Land Use = 45.0%	Land Use = 0.0%	Land Use = 45.0%
Slope = 33.3%	Slope = 45.0%	Slope = 45.0%	Slope = 0.0%
Soil = 33.3%	Soil = 0.0%	Soil = 45.0%	Soil = 45.0%

*Flood Risk Assessment Map

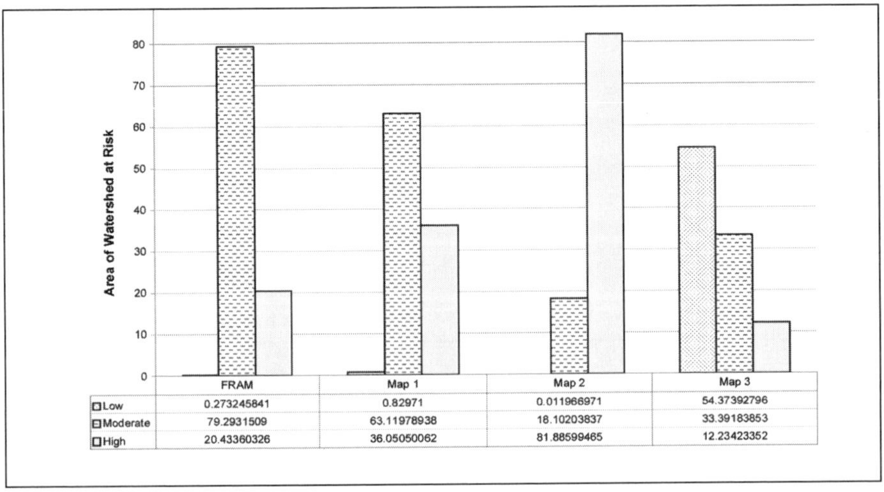

Figure 7.5 Comparison between FRAM and sensitivity maps.

Table 7.4 Polygon Count for each Risk Class from Sensitivity Maps

	FRAM*	Map 1	Map 2	Map 3
Flood Risk	**Polygon Count**	**Polygon Count**	**Polygon Count**	**Polygon Count**
Low	137	416	6	27,262
Moderate	39,756	31,647	9,076	16,742
High	10,245	18,075	41,056	6,134

*Flood Risk Assessment Map

Table 7.5 Percentage Change from Flood Risk Assessment Map

	Percentage Change from Flood Risk Assessment Map		
	Soil Removed	**Land Use Removed**	**Slope Removed**
Low Risk	0.0%	0.0%	+53.0%
Moderate Risk	-16.0%	-61.0%	-45.0%
High Risk	0.16%	+61.0%	-8.0%

From the above results, it can be seen that

1. If the land-use layer was not considered, there would be 61% more areas deemed high-risk areas than shown on the flood risk assessment map.
2. If slope layer was not considered, there would be a 53% increase in the number of areas that would be classified as low risk. This means that the land-use alteration would cause a generalization in high-risk areas, and the slope alteration would cause a generalization in low-risk areas.
3. Additionally, a change to the soil layer weighting resulted in more areas being classified within the high-risk group.
4. Based on the above statistics, it can be concluded that no one layer

should be excluded or under-weighted for the flood risk assessment map development. To do this would cause some areas that are high risk to be put into lesser hazard classes (that is, due to altering slope layer) or vice-versa (that is, due to altering land-use or soil layers).

7.5 Conclusions

The formulation of a tested flood risk assessment map and flood sensitivity maps using GIS technology and limited resources is an encouraging sign for the future of such work in Trinidad. Such maps, if utilized properly, will allow efficient, timely and cost-effective methods, to reduce flooding and flood-related losses and, in the long run, will assist with the proper management of all our watersheds. It should be noted that because the overlay operation is a relatively easy one to master, more than one scenario can be developed (sensitivity analysis), and after some time, the best input weights and ranking systems will be determined for the local environment. The cartographic model developed is a relatively simple one, and it identifies three main factors as potential flood causing agents: soil types, land use/land cover and slope. However, because of the versatility of GIS, additional parameters such as distance from the river in the form of buffers, geology or aspect can be added to the model. It should also be noted that the dynamic and unpredictable force of nature needs to somehow be factored into the entire equation. This type of holistic approach would mean that specialized stations that monitor current wind speeds, humidity, tidal levels, soil moisture, and so on would have to be in place, and this real-time data, in conjunction with the developed flood risk assessment maps, will have to be used to make decisions.

It can be seen that the way forward is to embrace this new technology. As previously stated, the absence of any real historical records and the generalized nature of existing flood records mean that local planning and disaster management authorities are responding in a reactive, rather than proactive, manner to floods. This can be seen as, year after year, piecemeal efforts are devised in an attempt to alleviate flooding. These projects include widening and dredging of rivers, fixing pumping stations, raising riverbanks and countless other proposals. However, these measures are short-term solutions and will not prevent the re-occurrence of floods. By utilizing the power of GIS, town and country planners can monitor the size and location of settlements, Ministry of Works officials can identify rivers that need work, disaster managers can specify areas that are high risk and develop engineering solutions to reduce potential flooding, and emergency officials can give accurate, specific and timely flood warnings to residents.

References

Adamson M. 2003. Floods: Causes, management and relief. Chartered Institution of Water and Environmental Managers, Ireland.

Allan, J.A., and M. Bradshaw. 1981. *Remote sensing in geological and terrain studies*. London: Remote Sensing Society.

Aziz, F., N. Tripathi, O. Mark and M. Kusangi. 2002. Dynamic flood warning system: An integrated approach to disaster mitigation in Bangladesh. Map Asia.

Baban, S.M.J., and R. Kantasingh. 2005. Mapping floods in the St Joseph Watershed, Trinidad, using GIS. *International Association of Hydrological Sciences (IAHS)*, pubiication no. 295:254–64.

Burrough, P.A., and R.A. McDonnell. 1998. *Principles of geographical information systems, spatial information systems and geostatistics*. Oxford: Oxford University Press.

Kantasingh, R. 2005. Managing floods in St Joseph River Catchment, Trinidad, using geoinformatics. MSc thesis, University of the West Indies, Trinidad and Tobago.

Chan, C., K.W. Howard, B.E. Vieux and J.E. Vieux. 2003. Operational deployment of a physics-based distributed rainfall-runoff model for flood forecasting in Taiwan. Paper presented at the IAHS General Assembly. Sapporo, Japan, 3–11 July.

Environmental Management Authority (EMA). 2001. *Initial national communication of the Republic of Trinidad and Tobago under the UN Framework Convention on Climate Change*. Port of Spain: Government Printery.

———. 1996. *State of the environment report: Trinidad and Tobago*. Port of Spain: Government Printery.

Forestry Report. 2001. *Squatting in forest reserves, prohibited areas and wildlife sanctuaries*. Port of Spain: Government Printery.

Grover, S. 1999. Perspectives of GIS modelling in hydrology. *GIS Development.*

International Tropical Timber Council (ITTC). 2003. Achieving the ITTO objective 2000 and sustainable forest management in Trinidad and Tobago. Thirty-fourth session. http://www.itto.or.jp/ittcdd_ses/thirty_fourth_sessions.html.

Land Settlement Agency (LSA). 2001. Trinidad and Tobago country report. Special session of the United Nations General Assembly for an Overall Review and Appraisal of the Implementation of the Habitat Agenda.

Langbein, W.B., and K.T. Iseri. 1995. *Science in your watershed: General introduction and hydrologic definitions.* Manual of Hydrology: Part 1. Geological Survey Water-supply Paper 1541-A. http://water.usgs.gov/wsc/glossary.html.

Leopold, L.B., and T. Maddock, Jr. 1954. *The Flood Control Controversy.* New York: Ronald Press.

Mitchell, A. 1999. *The ESRI guide to GIS analysis.* Vol. 1. Redlands, CA: ESRI Press.

Ramkisoon, K. 2000. An analysis of the legal framework for state land management in Trinidad and Tobago. Report for Land Use Policy and Administration Project (LUPAP).

Sanyal, J., and X.X. Lu. 2003. Application of GIS in flood hazard mapping: A case study of Gangetic West Bengal, India. Map Asia. *Proceedings from the Second Annual Asian Conference and Exhibition in the Field of GIS, GPS, Aerial Photography and Remote Sensing.*

Satelli, A., K. Chan and E.M. Scott. 2000. *Sensitivity analysis.* New York: John Wiley and Sons.

Seth, S.M., S.K. Jain and M.K. Jain. 1999. Remote sensing and GIS application studies at the National Institute of Hydrology. Map India. *Proceedings from the Second Annual Conference and Exhibition on GIS/GPS and Remote Sensing.*

Sharma, T., P.V. Satya Kiran, T.P. Singh, A.V. Trivedi and R.R. Navalgund. 2001. Hydrologic response of a watershed to land use changes: A remote sensing and GIS Approach. *International Journal of Remote Sensing* 22, no. 11: 2095–108.

Volusia. 2001. Volusia County Government GIS: http://www.volusia.org/gis/spatial.htm.

Ward, R.C., and M. Robinson. 2000. *Principles of hydrology.* London: McGraw-Hill.

World Bank Group. 2003. http://www.worldbank.org/data/countrydata/aag/tto_aag.pdf.

Zeiler, M. 1999. *Modelling our world.* Redlands, CA: ESRI Press.

CHAPTER 8

A New Examination of Floods in the Region

Debris Floods and Debris Flows in the Caribbean

RAFI AHMAD

Abstract

Flooding in the small and steep mountain watersheds of the Caribbean comprise a number of physical processes identified as common water floods, debris floods and debris flows. It is important to make a distinction between sediment (debris flows and debris floods) floods and water floods as this will have serious implications for planning and flood mitigation measures in the region. Sediment floods are commonly misidentified as water floods, and it has been shown that many of the recent disastrous flood events in the Caribbean were sediment flows rather than water floods.

The aim of this chapter is to use Jamaican examples to show that debris floods and debris flows are pervasive in small and steep channels throughout the Caribbean. In a majority of cases, it is the deposition of sediment rather than water that leads to death and destruction, for example the 2004 flood disaster in Haiti. Since debris floods transport large quantities of sediment, they are a distinct process compared to water floods. It is not uncommon, however, to observe the entire spectrum of flow behaviours in a particular channel: stream flow–hypoconcentrated flow–debris flow. However, the sediment flow processes are recorded and treated as water floods in the Caribbean and consequently, most mitigation is designed to contain water floods. This is unfortunate, because mitigation strategies for sediment flows are significantly different from those for water floods.

8.1 Introduction

Water-related disasters, often referred to as hydrogeologic disasters, are most common in the small island states of the Caribbean. In the hydrologic and flood management terminology, these processes are manifested as a number of flow types and rheological behaviours ranging from water floods to debris flows (Ahmad and Baban 2004).

What is commonly described as flooding in the small and steep mountain watersheds of the Caribbean, comprises a number of physical processes identified as common water floods (Newtonian fluids), debris floods (hypoconcentrated flows) and debris flows (visco-plastic behaviour). Debris flows and debris floods are commonly misidentified as water floods, and it has been shown that many of the recent disastrous flood events in the Caribbean were sediment flows rather than water floods as commonly assumed (Costa and Jarrett 1981; Ahmad 1991, 1995).

On the island of Jamaica, drainage areas of watersheds range between 232 km^2 to 1,892 km^2. According to the Water Resources Authority of Jamaica, the average rainfall in these basins varies from a low of 312 mm^3/year to a high of 5,068 mm^3/year, with corresponding surface runoff values at 81 mm^3/year to 2,452 mm^3/year. The main or trunk stream generally consists of third-order channels. However, head waters of these drainage basins comprise first- and second-order streams sub-drainage basins, with average sizes ranging between 10 ha to 500 ha hollows on slopes, which are old to young landslide scarps. These are referred to as small and steep drainage basins in this chapter.

Land use in Jamaica is changing rapidly in order to meet the economic and housing needs of the growing population. The high incidence of hydro-geological hazard events in the last decade is symptomatic of this change (Ahmad and Baban 2004). The frequency of sediment flood disasters in Jamaica, approximately one in every four years during the last 50 years, makes it abundantly clear that these events are widespread and costly (Table 8.1). Jamaica's vulnerability to multiple hazards is one of the main threats to the sustainable development of the country. However, this situation can be avoided since the causes of the majority of natural hazard events occurring in Jamaica are relatively well understood. It is possible to develop rainfall thresholds for initiation of debris flows to aid early warning and prediction of hydrogeologic hazard events with a reasonable degree of reliability; that is a fact that makes sediment floods amenable to measures directed at avoidance and/or prevention (Ahmad and Baban 2004).

In 2004 and 2005, sediment floods affected several Caribbean islands including Haiti, Dominica, Grenada, Jamaica, St Lucia, and Trinidad and Tobago. Sediment floods cause death and injury, as well as significant damage to lifeline structures, residential dwellings, water supply, road network, power lines, agriculture and businesses throughout the Caribbean region (Ahmad and Baban 2004). Although direct and indirect losses from recurring sediment disasters are in the order of millions of dollars for every event, for example Jamaica (Table 8.1), this subject has not received due attention from either the local hazard professional community or disaster management officials. Indeed, all sediment flows are recorded and treated as water floods, and most mitigation is designed to contain water floods. This is unfortunate because mitigation strategies for sediment flows are significantly different from those for water floods. Sediment floods are amenable to avoidance and correction, and economic losses may be significantly reduced.

It is important to make a distinction between sediment floods and water floods. Costa and Jarrett (1981) have shown that protective measures for water floods may not be effective for debris flows, and indirect-discharge estimates in sediment flow channels may not be accurate. Hydrological processes and channel dynamics in the relatively small and steep mountain watersheds of the Caribbean are not comparable to those operating in large river basins (Ahmad and Baban 2004). Since flooding processes are markedly different in the two environments, the response and management for two scenarios is also different. For example, in the Bybrook area of Portland parish it was noted that

Table 8.1 Damage to Infrastructure Caused by some of the Landslide and Flooding Events Occurring during 1986–1998 in Jamaica

Event	**Damage (estimated cost in US$)**
Flood rains May–June 1986	• Cost of repairs to damaged and destroyed road network, $16 million. • Village of Preston, St Mary, destroyed; 17 families displaced; replacement cost in 1986, $273,000.
Hurricane Gilbert rainfall 8–19 September 1988	• Approximately 60% of island's water facilities damaged; repair costs estimated at $10 million. • Boar River Water Supply Pipeline damaged. Repairs to island's road network estimated at $19.3 million. • 478 landslides along 108 km of roads in north-western St Andrew (4% of the island's total road system) blocked by landslides. • Landslides delivered an estimated 20,000 m^3 of sediment to rivers.
Flood rains 21–22 May 1991	• Island-wide damage estimated at $30 million. • Bog Walk Gorge road blocked by a landslide, forcing it to remain closed for more than six months.
Tropical storm Gordon rainfall 11–12 November 1994	• Approximately 241 km, or 2.3%, of island's total road network damaged; cost $2 million. • Damage to water systems estimated at $834,000.
Flood rains 3–4 January 1998 Portland	• Total damage approximately $8 million.
Period 1986–1998	• Total: Approximately $86.25 million.

Figure 8.1 An inadequate culvert design in the Bybrook area of Portland, Jamaica. Following 2001 rainfall associated with the Tropical Storm Michelle, many of culverts designed for water flood discharge failed to function as passages for debris floods and debris flows.

many of the culverts and bridges failed to function in the event of debris floods and debris flows since they were designed to accommodate water floods (Figure 8.1) and were thus too small.

The aim of this chapter is to explain that many of the recent disastrous flood events in the Caribbean were sediment flows rather than water floods. Furthermore, the chapter will use Jamaican examples to show that debris floods and debris flows are pervasive in small and steep channels throughout the Caribbean. Finally, to encourage the decision to make the necessary provisions for sediment flows as in a majority of cases, it will be demonstrated that it is the deposition of sediment rather than water that leads to death and destruction.

8.2 Sediment-water Flow Types

Costa and Jarrett (1981), Costa (1984), and Hungr et al. (2001) have described the physical geomorphology, material properties, rheology,

Table 8.2 Sediment-water Flow Types in Channels, Modified From Costa, (1984)

Flow Type:	**Water Flood Stream Flow**	**Debris Flood Mud Flood Hypoconcentrated Flow**	**Flow Debris Flow**
Sediment load by weight	1–40%	40–70%	70–90%
Bulk density g/cm^3	1.0–1.3	1.3–1.8	10–20
Strength dynes/m^2	<100	100–200	>200
Fluid type	Newtonian	Approximately Newtonian	Visco-plastic
Deposits and landform	Sorted, stratified sheets and bars	Poorly sorted, weakly stratified, no sharply defined margins	Levees and lobes of very poorly sorted, largely unstratified debris; large clasts on top and at face of lobe

differentiation of water floods and debris flows, and the classification of sediment-water flow types (Table 8.2).

Debris flows may be regarded as gravity-induced mass movement comprising poorly sorted rock debris, 70–90% by weight, and a process intermediate between landsliding and water-flooding. Its flow properties vary, depending upon the sediment size, varying between clay to boulders, sorting and water content. Debris flows are highly viscous and are therefore able to transport large rock blocks (> 1 m) on their surface. Their erosive power is several orders of magnitude higher than water floods and hypoconcentrated flows (Ahmad and Baban 2004). Following Hungr et al. (2001), "debris flood is a very rapid, surging flow of water, heavily charged with debris, in a steep channel".

Many debris flows become diluted with water downstream to become debris floods. It is regarded as a mass transport phenomenon distinct from landslides and simulating what have been described as hypoconcentrated flows or sediment slurries, which can easily move on gentle slopes (Costa 1984; Hungr et al. 2001).

Figure 8.2 A van and houses engulfed by debris flood deposits in 2001 on the Bull Bay River Debris Fan, St Andrew, Jamaica. Note that the vehicle and houses are undamaged.

Figure 8.3 Debris flow deposit in Bybrook triggered by rainfall associated with Tropical Storm Michelle, 2001. A car engulfed by debris is damaged. Note poor sorting and lack of stratification in the deposit.

Figure 8.4 Waterflooding at the entrance of the Mona campus, University of the West Indies, Hurricane Ivan rainfall, September 2004.

Figure 8.2 shows a classic case of a debris flood from Bull Bay in Jamaica where a motor van is surrounded by debris, but window glass remains damaged. In contrast, a debris flow illustrated in Figure 8.3 from Bybrook, Jamaica shows a motor car damaged and partly buried by a debris flow deposit. In contrast, water floods are turbulent flows carrying relatively small amounts of sediments (1–40% by weight) and having low densities, with generally stratified and well sorted deposits (Figure 8.4). These features help to map paleo-flow types in channels.

8.3 Study Area

Jamaica is located in the track of Atlantic hurricanes and also within a seismically active plate boundary zone. Geologically young landforms with steep hillsides are characteristic. Most of the population centres in the Caribbean are sited on alluvial fans, which are created by deposition of eroded sediments from uplands. These fans are often located at the base of faulted mountain fronts, for example, the Liguanea debris fan, which hosts the greater part of Kingston, the capital city of Jamaica. The town of Harbour View is also sited on the alluvial fan of the Hope River. Alluvian fans are sites of destructive sediment floods. The alluvial fan floods are characterized by relatively shallow depths and tend to carry a large volume of sediments (Ahmad and Baban 2004). In the Caribbean, most coastal fan surfaces are highly urbanized regions, and the gentle topography on fans are the favoured building sites. Alluvial fans and alluvial fan flooding in the Caribbean show great diversity because of the evolutionary history of fans, rates and styles of recent tectonics, source area rock types, vegetation conversion and land use (Ahmad and Baban 2004).

Hazards that affect the island most frequently are geohazards, landslides and flooding. The high incidence of these hazard events in Jamaica is a result of combinations of geological, geophysical and geographic factors. These hazards are important because of their frequency, associated loss of life, disruption of socioeconomic activities, and effect on the built and natural environment. Landslides account for most of the natural disasters that have occurred on the island during the last decade, and they continue to present risks to life and property. The high incidence of landslides and flooding are also symptomatic of changing land use. Landslides and floods are a recurring cause of death and injury and have damaged and destroyed rural settlements, schools, public and private property, roads, bridges, culverts, retaining walls, agricultural lands and crops, water pipelines, electricity transmission lines, and submarine cables. In addition, slopes that have been denuded by landslides suffer from accelerated soil erosion. This means that the indirect economic costs of such natural disasters, therefore, can be several orders of magnitude higher than the direct costs (Ahmad et al. 2004).

8.3.1 Landslides

The controlling factors and mechanisms that favour the occurrence of landslides in Jamaica are well known (Ahmad 1995). Triggering mechanisms include earthquakes and/or heavy rain. Earthquakes with magnitudes 4.5 or greater on the Richter scale have caused landslides. In addition, 200 mm to 300 mm of rain falling within a 48-hour period also initiates shallow landslides that are quickly transformed into debris flows and mud flows that cause extensive damage. Landslides are also strongly influenced by rock type and geological structures, with the density of faults and joints being very important factors. In many cases, damage from landslides has incorrectly been ascribed to damage from floods (Ahmad and Baban 1995). In order to reduce landslide-related losses, this perception must change within the planning and mitigation community.

Flash flooding caused by landslide dam failure is a significant hazard in Jamaica and the Caribbean, and it is common in the mountainous areas (Robinson et al. 1996). Large volumes of sediments generated by landslides and debris flows occasionally create ephemeral dams across river courses in Jamaica, which result in the impoundment of large volumes of water. This phenomenon is widespread in the Jamaican river valleys, where there are historical accounts and also geomorphic evidence of large rock and earth slumps and slides, debris and mudflows, and rock and debris avalanches, which created landslide dams. Subsequent overtopping and/or or breaking through of the temporary landslide dams have resulted in landslide and debris outburst floods. For example, this occurred in the 1937 Millbank flood disaster on the Rio Grande and Swift River in the parish of Portland.

Sediment floods have, in many instances, led to deposition of huge amounts of debris into the rivers with the consequent elevation of the riverbed, for example, the Swift River, following rainfall associated with Tropical Storm Michelle in 2001.

The rainstorm of 3–4 January 1998 triggered widespread debris flows and mud flows that choked the water courses, thus aiding flooding in the Rio Grande Valley, Portland, in the northeast of the island. This disaster resulted in direct losses estimated by the National Environmental and Planning Agency and the Office of Disaster Preparedness and Emergency Management at some US$8 million. In the Grants Level area of Portland, a debris flow killed four people and left several others seriously injured (Ahmad and Baban 2004).

8.4 Case Study: St Elizabeth and Westmoreland, Jamaica, June 1979

The following summary of debris flows and flooding triggered by the flood rains of 12 June 1979, in the parishes of St Elizabeth and Westmoreland, western Jamaica, is based on the section on natural hazards prepared by the author in Halcrow and Partners (1998). On the night of 12 June 1979, a tropical depression dropped about 865 mm of rainfall in 8 to 10 hours, affecting an area of some 2,500 km^2 in western Jamaica, leading to extensive flooding and debris flows. The highest point rainfall was estimated at 864 mm. Approximately 1,150 population centres, involving some 160,000 people, were affected, and 44 people were killed. The damage to infrastructure and housing was estimated at J$70 million.

8.4.1 Debris Flows and Debris Floods

Flooding and debris flows disrupted communication, and houses were swept away. Many houses, roads and agricultural fields were buried under the landslide debris. Figure 8.5 shows a house near Bluefields, completely overwhelmed by limestone debris. Slopes in the limestone highlands, with angles at 20° to 40°, are generally deforested and characterized by short and steep dry gullies. Loose sediments (colluvium) provide a soil cover. These sediments were mobilized by the heavy rainfall as debris floods.

Landslides were common within the hilly terrain, comprising fine and coarse volcaniclastics including shale, mudstone and sandstone. Landslide debris moved along gullies, which acted as chutes and were deposited as fans, lobes and ridges over much of the lowland areas adjoining the hills and mountains and coastline (Figure 8.6). Landslide debris choked many of the rivers and, consequently, rivers overflowed their banks. The level of flooding observed was not simply a function of excessive surface runoff, which the river channels were unable to carry, but also was related to a reduction of channel cross sections as a consequence of sediment deposition. The road systems often served as major conduits for the runoff and, as a consequence, were severely scoured and damaged.

Figure 8.5 A house in Bluefields completely buried as a result of debris flow; 1979 flood rains in western Jamaica. (Courtesy of the Jamaica Information Service, Photo Library.)

Figure 8.6 Sediment surges in the first order channels created a sediment delta at the mouth of the Bluefields River; 1979 flood rains western Jamaica (Jones 1981).

In the alluvial lowlands, flooding was extensive and prolonged as in the Cabarita River system. The blockage of bridges and culverts by sediments and organic debris led to rivers overtopping their channels and breaching roads, embankments and retaining walls at various places including Ricketts River area and Bluefields. According to O'Hara (1990), "the most spectacular damage occurred at junctions of the plains with steeply rising White Limestone hills". The settlements at Cave in Surinam Quarters and Anglesea to Whithorn area including Waterworks, at the eastern edge of Georges Plain, were severely damaged as a result of the combined action of flooding and debris flows. At these sites, landslide debris destroyed houses, roads and agricultural lands, often burying them beneath a layer of boulder deposits reaching a thickness of some 4 m.

In summary, a hilly terrain, underlaid by highly fractured limestone bedrock and colluvium-covered slopes, together with heavy rainfall, promotes a natural propensity to landslides (debris flows). Also, there have been many failures on road cuts, retaining walls and fills. Massive surges of sediments were transported as debris floods. Flooding and landslides are known to occur at the same time and under the same weather conditions, and they are often interrelated. The drainage was choked with landslide debris and organic material, which diminished the effectiveness of channels to discharge runoff.

8.5 Discussion

Geologically young landscapes of the Caribbean are particularly vulnerable to debris flows and debris floods. The islands are mountainous; therefore, most settlements and much of the key infrastructure are located along the coast and/or on relatively flat alluvial and debris fans at the mouths of the streams. Potentially hazardous areas close to major population centres are being increasingly used for housing development. High annual precipitation and periodic short-duration/high-intensity rainfall from tropical storms and hurricanes trigger landslides in small and steep drainage basins of the Caribbean. Debris flows, mud flows and debris floods are frequent, following significant rainfall events. Sources of sediments for debris floods include rainfall-induced slope failures, loose sediments on slopes derived from old landslides, and sediments in the water courses related to previous sediment-water floods. Mobilization potential of these sediments is rather high. Rainfall thresholds that may mobilize sediment-water floods need not be associated with hurricane rainfall. It is not uncommon to observe the entire spectrum of flow behaviours in a channel: stream flow–hypoconcentrated flow–debris flow.

Although landslides are a serious geological hazard on most of the Caribbean islands, being part of an erosive process they also provide rock debris that accumulates at the base of hill slopes as debris fans and in water courses as channel deposits, where it is mined as an aggregate material for the construction industry. Landslide debris is therefore an important economic resource. In Jamaica, there are several aggregate operations; for example, Morant River, Rio Minho, Yallahs River and Wagwater River have created business and employment opportunities and sustain the livelihoods of many persons in the rural sections of the island.

8.6 Conclusions

An approach for sediment-water floods for the small island developing states (SIDS) was advanced. Debris flows, mud flows and debris floods, collectively designated as sediment-water flows, frequently occur in small and steep drainage basins following significant rainfall events. Sources of sediments include storm-induced landslides, loose sediments on slopes derived from old slope failures and sediments in the watercourses related to previous sediment-water floods. The mobilization potential of these sediments is high and, in many cases, it is the deposition of sediments rather than water that leads to the hazardous situations. The sources of sediments or the rock debris in the small and steep drainage basins are the slope movements, landslides and debris flows that have occurred in the past and are likely to continue into the future.

Given the relatively small size of the watersheds in the Caribbean islands, it is important to distinguish different flow types to adopt appropriate mitigation. Failure to do so often results in losses that are easily avoided and managed. A post-disaster damage assessment survey in the parish of Portland, in the wake of rainfall associated with Tropical Storm Michelle in 2001, and later events in other areas, revealed many such cases. For example, in the Bybrook area of Portland parish it was noted that many of the culverts and bridges failed to function as debris flow chutes since they were small and designed to accommodate water floods.

Landslide debris is an important economic resource that has created business and employment opportunities, and it sustains the livelihoods of many rural communities throughout SIDS. A plan for "debris management", however, should be an essential component of the sediment-water flow mitigation.

References

Ahmad, R. 1991. Landslides triggered by the rainstorm of May 21–22, 1991. *Jamaican Journal of Science and Technology* 2:1–13.

Ahmad, R. 1995. Landslides in Jamaica: Extent, significance, and geological zonation. In *Environment and development in small island states*, ed. D. Barker and D.F.M. McGregor, 147–69. Kingston: The Press, University of the West Indies.

Ahmad, R., and S.M.J. Baban. 2004. Sediment water floods in the Caribbean. Paper presented at the workshop Enduring Geohazards (Landslides and Floods) in the Caribbean Region, Learning Resource Centre, University of the West Indies, St Augustine, Trinidad. 8 December.

Ahmad R., S.M.J. Baban, K. Sant and A. Chimchame. 2004. Flooding and landslides in the West Indies: Digging deeper into the dirt. *The Hammer* (March): 9–13.

Costa, J.E. 1984. Physical geomorphology of debris flows. In *Developments and applications of geomorphology*, ed. J.E. Costa and P.J. Fleisher, 268–317. Berlin: Springer-Verlag.

Costa, J.E., and R.D. Jarrett. 1981. Debris flows in small mountain stream channels of Colorado and their hydrologic implications. *Bulletin of the Association of Engineering Geologists* 18:309–22.

Halcrow and Partners. 1998. Geology and natural hazards. Technical Report no. 4, Multi-sectoral Pre-investment Programme, South Coast Sustainable Development Study for Government of Jamaica.

Hungr, O., S.G. Evans, M.J. Bovis, and J.N. Hutchinson. 2001. A review of the classification of landslides of the flow type. *Environment and Engineering Geoscience* 7:221–38.

Jones, E. 1981. Geomorphological aspects of 1979 flood rains. *Journal of the Geological Society of Jamaica* 20:53.

O'Hara, M. 1990. Flood hydrology of western Jamaica: A study in a karstic limestone environment. *Singapore Journal of Tropical Geography* 11:100–116.

Robinson, E., R. Ahmad, C. Phillip-Jordan and M. Armstrong. 1996. The Burlington landslide: An example of an ancient landslide dam, Rio Grande, Jamaica. *Journal Geological Society of Jamaica* 31:37–42.

CHAPTER 9

Mapping Flood-prone Areas

A Geoinformatics Approach

SERWAN M.J. BABAN and FRANCIS CANNISUS

Abstract

Flooding is the most common hazard that affects Caribbean territories. It is influenced mostly by heavy rainfall, development, land-use pattern and the geomorphological properties of hydrological catchments.

This chapter develops a methodology to identify flood-prone areas in Trinidad, using a variety of sources including available flood maps, topographic maps, aerial photos, digital elevation models (DEM), newspaper articles and other historical data. Based on available data, 106 flood events were identified in Trinidad from 1986 to 2006. These events were analysed and related to geographical locations. The areas that were repeatedly flooded during this 20-year period were identified and mapped. The geophysical terrain characteristics such as slope, elevation, geology and rainfall for these susceptible areas for flooding were derived. These terrain characteristics were then used to identify potential flood-prone areas and to generate a map identifying these areas in Trinidad. The developed methodology is simple and easy to implement. The outcome map is useful for hazard identification, the regulation of future development and the establishment of flood insurance premium rates.

9.1 Introduction

Flooding is the most common disastrous act of nature among all catastrophes leading to economic losses and death (Sharma and Priya 2001). A flood can be defined as a mass of water, which produces runoff on land that is not normally covered by water or, alternatively, a flood is a fairly high flow, which overburdens the natural channel provided for the runoff (Ward 1978; Cooke and Doornkamp 1974).

Heavy rains in some hydrological catchments tend to cause floods. Rainwater overflows the riverbanks, causing inundation in low-lying areas along the river and in the downstream areas (Cooke and Doornkamp 1974). However, the actual impact of flood events depends mainly on the physical characteristics and the conditions of hydrological catchments. For example, if the topography of the catchment is steep, the velocity of the floodwater will be great, thus causing destructive damages, though the inundation areas would be limited and the duration short. If the topography of the catchment is gentle, the flood will be extensive and will last for a long period. Furthermore, if the vegetation cover and distribution within a catchment is poor, and the geology is fragile, the flood will carry and contain large amounts of debris and will be more destructive (Few et al. 2004).

Since humans historically have established settlements in river valleys, floods have created hazards for human communities for centuries (Wohl 2000). In turn, human activities involving environmental degradation, deforestation and inappropriate land use often encourage flooding. Riverine flood waters often carry a considerable amount of sand, silt and debris that can block channels and dams, intensifying flooding upstream. In the Caribbean, floods can also result in secondary hazards such as mudslides and landslides, which can destroy lives and properties. In general, there are two main causes of floods: natural causes and anthropogenic causes (Sharma and Priya 2001). The natural causes of flooding include atmospheric hazards (heavy rainfall, El Niño, storm surges), tectonic hazards (landslides, tsunamis) and technologic hazards (dam failures) (Ward 1978), while anthropogenic causes of flooding include extensive development of flood plain areas (urbanization, deforestation), poor farming (over-grazing, over-cultivation) and poor water management (Smith 1991; Frasier 2005)

A flood-prone area can be described as an area that has the potential to flood and may or may not be within a water body's regular flood plain (Cooke and Doornkamp 1974). It is evident that most of the extensive flood-prone areas are located along the coastal plains and riverine areas, which tend to coincide with densely populated and highly built-up areas (Chan 1997). Literature (Cooke and Doornkamp 1974; Smith 1991) indicates that the following are the most vulnerable landscape settings for floods:

1. Flat and low lying or areas with gentle slopes with poor drainage in their natural state. These settings will suffer the most frequent flooding.
2. Low-lying coasts, deltas and estuarine areas. These areas are often exposed to a combined threat of floods from rivers and high tides.
3. Small catchment basin, basins characterized by a combination of steep topography, little vegetation and heavily developed urban settings.
4. Areas below unsafe or inadequate dams.
5. Low-lying inland shorelines.
6. Catchments with rivers functioning with reduced carrying capacity and flow constraints due to vegetation, tidal influences or infra structure such as bridges and culverts.
7. Watersheds with short longitudinal axes. The time of arrival of a flood wave is generally shorter than for equivalent watersheds having longer longitudinal axes.
8. Watersheds characterized by high runoff, if the surrounding land has a high runoff potential, due to development of impervious soils.
9. Alluvial fans, which tend to have a history of flooding and often provide attractive development sites due to their commanding views and good local drainage.

Maps of flood-prone areas can be used for flood-hazard identification, regulation of future development, helping communities to understand where flood-prone areas are located, establishing flood insurance premium rates, and identifying areas having unique, natural and beneficial functions (Jones 2004). However, mapping of flood-prone areas requires considerable collection of historical data, accurate digital elevation data, discharge data and a number of cross-sections located throughout the watershed.

Spatial information technologies in the form of geographic information systems (GIS) and remote sensing, have made major contributions to flood management studies and natural hazard and risk zonation mapping (Billa et al. 2004). By using GIS, one can organize spatial data, visualize complex spatial relationships of a particular location make a spatial query about the characteristics of a particular location, and combine data sets from different sources (Baban and Kantasingh 2005).

In the context of flood management, GIS provides a broad range of tools for determining areas affected by floods and predicting areas likely to be flooded due to high discharge of rivers. GIS can also be used to create interactive map overlays, which can illustrate which area of a community may be in danger of flooding, thus coordinating mitigation efforts before an event and recovery after the event. Furthermore, the ability of GIS to develop three-dimensional topographical mapping and terrain modelling in the form of digital elevation models (DEM) is particularly useful for flood analysis and estimation (Jones et al. 1998). This chapter aims to identify flood-prone areas in Trinidad using a variety of sources, including available flood maps, topographic maps, aerial photos, DEM, newspaper articles and historical data. Once the flood events are identified, they will be related to specific geographical locations. The geophysical terrain characteristics such as slope, elevation, and rainfall for these susceptible areas for flooding will be deduced. These terrain characteristics will then be used to identify potential flood-prone areas and to generate a map identifying these areas in Trinidad.

9.2 The Study Area

Trinidad is a tropical island approximately 4,800 km^2, located in the southern West Indies, between the Caribbean Sea and the North Atlantic Ocean, northeast of Venezuela. It is one of the two main islands forming the Republic of Trinidad and Tobago. The island's most prominent natural feature comprises the three mountain ranges: Northern, Central and Southern Ranges that run east to west (EMA

1996). Between these ranges are broad plains, while extensive swamps cover the east, south and west coasts. Trinidad has numerous rivers, of which the Ortoire and Caroni Rivers are the longest. The development is mostly confined to the northwest, along an east–west dual carriageway, and to an industrial site located in the mid-west of the island.

Trinidad has a tropical climate influenced and modified by the surrounding sea and trade winds. The climate is characterized by a dry season from January to May and a wet season from June to December; the warmest month is July and the coolest is January. Rainfall varies annually from about 3,500 mm to 1,200 mm (Kenny et al. 1997). In terms of population distribution, the largest concentrations of population are located along the northern east-west and western north-south areas of the island (Ramroop 2005). Historically, most Indo-Trinidadian villages have been established in the flatlands of the central and the southern parts of Trinidad, which are mainly located on the flood plains (James 2004).

Trinidad has recently been experiencing an increase in flood events that have resulted in the inevitable flooding of low-lying areas and an increase of landslides throughout the country (Khan 2005). Flooding events seems to be influenced by development and changes in rainfall patterns in Trinidad (Baban and Kantasingh 2005). Flooding occurs from a range of causes and conditions in Trinidad. Chief among these is the heightened pace of urban development. This places considerable stress on the environment, more specifically:

1. quarrying activities contribute significantly to the siltation of riverbeds, thereby physically reducing the amount of water that the river retains;
2. slash-and-burn agriculture places undue stress on the land, eventually reducing its fertility;
3. unplanned housing developments on hills severely reduce vegetative cover, which often makes the difference between flooding and the natural runoff of water into traditional watercourses;
4. dumping everything from cars to household refuse in rivers and drains reduces the capacity of the drainage system, which leads to overflowing; and
5. extending urban development into flood plains has forced surplus waters to find other outlets.

In addition, since Trinidad is located in a tropical climate zone, it is prone to what is described as high-intensity rain. This means very heavy rain over a short period of time, resulting in flash flooding or heavy rainfall which results in the saturation of the soil. Thus, there is a critical need for maps of flood-prone areas.

9.3 Methodology and Data Development

Flood events and their locations during the past 20 years (1986–2006) were identified and collected from newspaper articles, the Water and Sewerage Authority, and the Office for Disaster Preparation and Management in Trinidad. Other available ancillary data for this project also included a DEM of Trinidad, a geo-referenced topographical map of scale 1:25,000, and maps of roads, rivers, soil and geology (Table 9.1).

9.3.1 Creation of the Flood-prone-location Map

In order to create a map of national flood-prone locations, 106 flood events were identified in Trinidad from 1986 to 2006. These events were analysed, related to geographical locations and the areas that were

Table 9.1 Type, Source and Scale of Available Ancillary Data Sets

Data Layer	Source	Scale
Coastline	Paper map	1:25,000
Contour	Paper map	1:25,000
Geology	Paper map	1:100,000
Soil	Paper map	1:150,000
Watershed	Vector data	–
River	Paper map	1:25,000
Road	Paper map	1:25,000
Land use/cover	Classified image	30m resolution
Flood	Paper map	1:150,000

Table 9.2 Areas Prone to Flooding in Trinidad

Communities	Biche, Caparo, Cedar Grove, Cocoyea Village, El Carmen, Felicity, Flanagin Town, Fonrose, Frederick Settlement, Independence Square, Kelly Village, La Paille, Lazarri Village, Mamoral, Manuel Congo, Marabella, Maraval, Monkey Town, Naivetė, Ortoire Village, Palmiste, Plaisance Village, Ravine Sable, Santa Phillipa, Solomon Road, South Quay, Spring Village, St Helena, Sumanie Trace, Warrenville
Roads	Charlotte Street, Charran Maharaj Trace, Chrysostom Trace, Mafeking, Clarke Road, Corner Of Penitence and Chacon Street, Dookie Singh Street, Edward Street, Freeling Street, French Street, Gill Street, Henry Street, Katwaroo Trace, Lower High Street, Mafeking Road, Mayaro Main Road Oropouche Trace, Poole Valley Road, Ramsamooj Street, River Side Road, Rochard Douglas Road, Rookminia Trace, Saddle Road, Sumanie Trace, Todds Road, Tragarete Road, Upper La Seiva Road, Warren Road, Winston Mahabir Street, Pleasantville Wrightson Road
Areas	Maraval, Port of Spain, Santa Cruz, Maracas, Arima, Cunupia, Caparo, South Oropouche, Cunapo, North Oropouche, Caroni

repeatedly flooded during this 20-year period were identified and mapped. The geophysical terrain characteristics such as slope, elevation, geology, lithology, soil texture, soil permeability and rainfall for these susceptible areas for flooding were derived. These terrain characteristics were then used to identify potential flood-prone areas, and to generate a map identifying these areas in Trinidad. These areas were then listed based on flooded communities, roads and named areas (Table 9.2).

9.3.2 Identifying and Mapping Supporting Factors for Flooding

The detailed examination of frequently flooded locations revealed that, in general, areas with gentle slopes, low elevation, geology and soil type watersheds with specific characteristics tend to be flood-prone areas in Trinidad. In terms of physical characteristics, these locations tend to comply with identified characteristics for flood-prone areas as described earlier in section 9.1. More specifically, the following physical attributes were deduced from areas subjected to floods in Trinidad over the last 20 years:

Slope: Flood-prone areas are normally flat or with very gentle slope. In Trinidad, slope was found to be the most influencing characteristic as more than 95% (101 locations) of the identified flood locations were located with < 1% slope (Table 9.3).

Elevation: Areas with low elevation tend to suffer the most frequent flooding as they are often characterized by low drainage and high sediment deposits, and they are located at the downstream area of watersheds. The analysis shows that out of 106 collected locations in Trinidad, 94 of them, that is, 89% of the locations, are located in areas with less than 50 m elevation (Table 9.4).

Geology: Geology can have a serious influence on flooding. The developed data set of floods indicate that 80% (85 locations) of the flooded areas are located within geological formations dominated by sand and gravel (Table 9.5).

Soil: Clearly, soil type can influence local drainage. For example, soil types with low infiltration rates, such as clay soil or shallow soil with hardpan near the surface, tend to sustain both frequent and long duration flooding. The analysis shows that 66% (70 locations) of the flooded areas are located on soil with very low infiltration rates (Table 9.6).

Rainfall: High-intensity rainfall upstream tends to cause flooding downstream. The analysis supports this concept, since most of the identified locations are located in low-rainfall areas (Table 9.7) and floods seem to be caused due to rainfall in the upper sections of the watershed. The analysis also indicates that frequently flooded areas tend to be located in watersheds with configurations that support concentrating runoff from the surrounding lands at the flood plain.

Table 9.8 shows the association between the number of frequently flooded areas and number of watersheds. This shows that 35 of 54

Table 9.3 Flood Locations in Range of Slope

Slope (%)	Flood Locations
1	101
2	1
3	3
4	1

Table 9.4 Flood Locations in Elevation Range

Elevation (m)	Flood Locations
0–10	32
10–20	27
20–30	8
30–40	15
40–50	12
50–60	2
60–70	3
70–80	2
80–90	1
90–100	4

Table 9.5 Flood Locations in Geology Types

Geology Flood Locations	
Volcanic	1
Shale	1
Sand and Gravel	85
Slate, Siltstone, Sandstone	1
Shale, Slate	1
Sand, Mudstone	5
Clay	2
Sandstone	4
Clay, Sand	2
Shale	1
Clay, Marl	2
Marl, Clay	1

Table 9.6 Flood Locations in Soil Types

Hydrologic Soil Group		Flood Locations
A	High infiltration rate: usually deep, well-drained sands and gravels with little silt or clay	2
B	Moderate infiltration rate: fine or moderate textured, well-structured soil, light sandy loams, silty loams	20
C	Below average infiltration rate; moderate- to fine-textured, shallow soil, clay loams	14
D	Very slow infiltration rate; usually clay soils or shallow soil with hardpan near the surface	70

Table 9.7 Flood Locations and Rainfall

Rainfall	Flood Locations
>3,500	0
3,250–3,50	0
3,000–3,250	1
2,750–3,000	6
2,500–2,750	11
2,250–2,500	7
2,000–2,250	15
1,750–2,000	25
1,500–1,750	31
<1,500	10
Total flood locations	106

Table 9.8 Flood Locations in Watersheds

Number of Watersheds	Flood Locations
1	12
1	9
1	8
1	7
1	6
1	5
6	4
4	3
4	2
15	1
19	0

watersheds experienced at least one-time flooding and are susceptible for flooding. This may change with the watershed condition such as change in vegetation cover and human settlements.

9.3.3 Developing Maps of Flood-Prone Susceptible Areas

There are several factors that are related to the site-specific ground conditions that may contribute to flood occurrence. The uses of geoinformatics in land suitability mapping is an approach of particular interest in locating flood-prone areas as it can be used to identify the most likely areas to flood based on the deductive approach using the incidence of past floods.

Lyle and Stutz (1983) have developed an automated land suitability mapping system, which involves the mathematical combination of three different factors: location, development actions and environmental effects. Information on the site requirements of the development and the effect of the development on the environmental quality of the site, along with an inventory of the site, are required. This information is then combined in a weighted overlay process. The weightings, or the amount of influence that a given factor will have on the final outcome, are determined by the construction of matrices. The final output is a computerized map showing the most and least suitable locations for the development action. This map can then be used to help support the planning decision. Indeed, such a technique was used in Manitoba, Canada, to explain to the public the stages that planners went through and the criteria they used to locate a toxic waste dump (Heywood 1991). Similar methods have also been used to assess site suitability for land application of waste in Vermont (Buckley and Hendrix 1986; Baban and Flannagan 1998) and for siting New York sludge management facilities (Gyulavary and Lopez 1991). In these examples, the siting criteria included social, economic and environmental factors.

Jensen (1986) offers a geoinformatics methodology for siting a waste disposal facility using Boolean logic functions. The final result is a map created by overlaying the individual environmental and cultural maps. Another method for siting a landfill waste disposal facility using geoinformatics was forwarded by Bonham-Carter (1994). The suitability of a site is determined by the overlaying of a number of maps containing environmental, social and economic data that relate to each constraint. As with the Jensen and Christensen method, Boolean logic is used to produce an output map showing which areas either are or are not suitable for the development action.

Given the constraints for siting flood-prone/susceptible areas, the task of finding the most and least susceptible areas for flooding can be achieved by the summing of attribute scores by map layer weights on a cell-by-cell basis using overlay functions within a GIS. In order for the output map to be meaningful and consistent, map weights had to add up to 100%, and the attribute scores had to be chosen using a scheme that was the same for each map (Baban and Flannagan 1998). In this study, overlay analysis was performed using the thematic layers of slope, elevation, geology, soil and watersheds. Slope acquires the highest score and watershed obtains the lowest score.

Figure 9.1 shows the developed procedure. The slope grid from DEM was classified into two classes. The first class was greater than 1% slope and represents areas not prone to flooding. The second class is less than 1% slope and contains areas susceptible to flooding. Areas

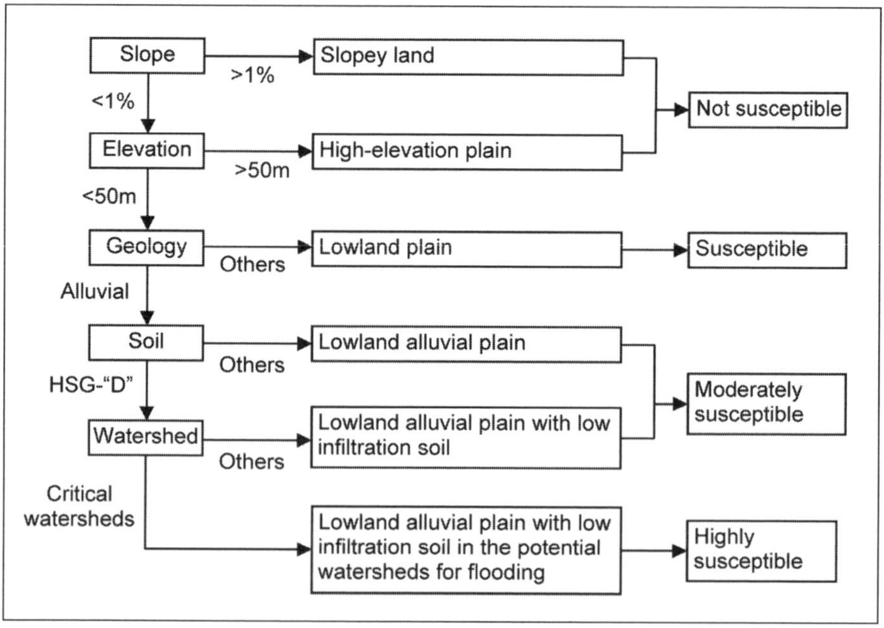

Figure 9.1 Developed procedure to identify and map flood-prone areas.

within the flat lands were then subdivided into two classes using the 50 m elevation threshold, derived from DEM, into "high-elevation plain" and "lowland plain". The geological formations in the identified areas were dominated by sand and gravel in low areas that were separated from the rest. Similarly, soil with a low infiltration rate was separated from "lowland alluvial plain". Based on the assumption that the watershed with at least one-time flooding indicates the high possibility of another flood event, the final classes were assigned as "lowland alluvial plain with low-infiltration soil" and "lowland alluvial plain with low-infiltration soil in the potential watersheds for flooding" (Figure 9.1).

9.4 Results

Flood mapping can be "backward looking" (showing past incidences of flooding) or "forward looking" (showing the probability of flooding irrespective of any past incidences of inundation) (JBA 2004). In this chapter, the flood location map is showing the past events, and the map of flood-prone areas, which was developed based on the geophysical characteristics, shows the susceptible areas of flooding.

9.4.1 Map of Flood-Prone Locations

Flood-prone locations (106) of the flood events from the past 20 years, frequencies and the distribution of flooding within the hydrological regions in Trinidad are shown in Figure 9.2. About 35% of the collected locations are in the Western Peninsula, where the Caroni River passes through intense settlements. The Southern Range, Cedros

Table 9.9 Flood Occurrence and the Number of Locations

Flood Occurrence within Last 20 Years	Flood Locations
1	63
2	21
3	12
4–9	6
10–13	4

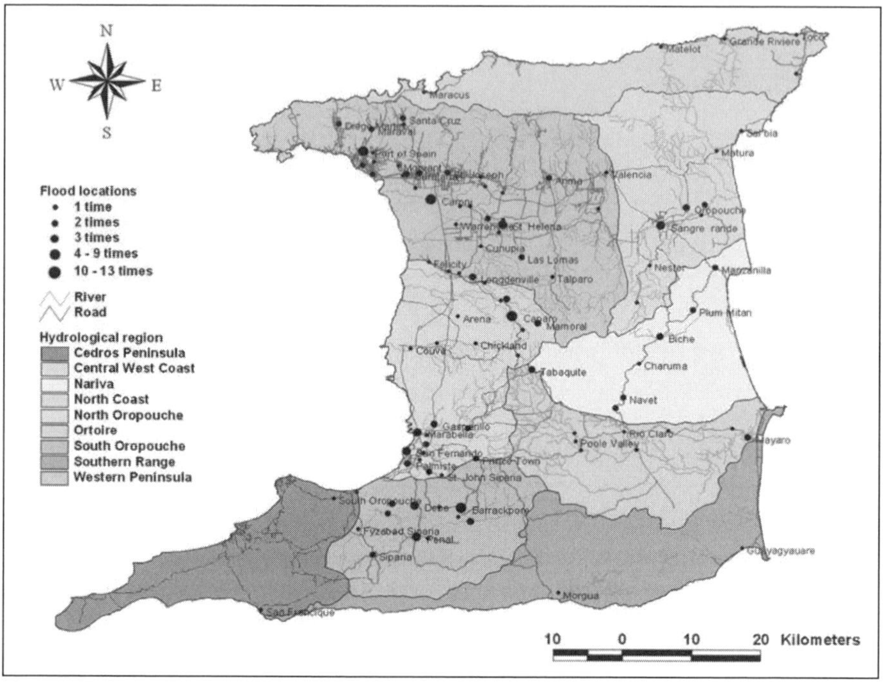

Figure 9.2 Flood-prone locations and frequencies within various hydrological catchments in Trinidad.

Peninsula and North Coast appear to be less susceptible hydrological regions with fewer than five identified flood locations.

The map shows flood locations and the frequency of flooding, therefore facilitating the spatial distribution of flood-prone areas nationally, as well as with respect to the frequency of flooding in specific locations (Figure 9.2). Table 9.9 summarizes the data and indicates that, during the last 20 years, floods have occurred more than 10 times in four locations, while flooding only occurred once in about 60% of the identified locations.

9.4.2 Terrain Liable for Flooding Based on Supporting Factors for Floods

Figure 9.3, obtained by overlaying various thematic layers in a GIS environment, exhibits the defined pattern of terrain characteristics

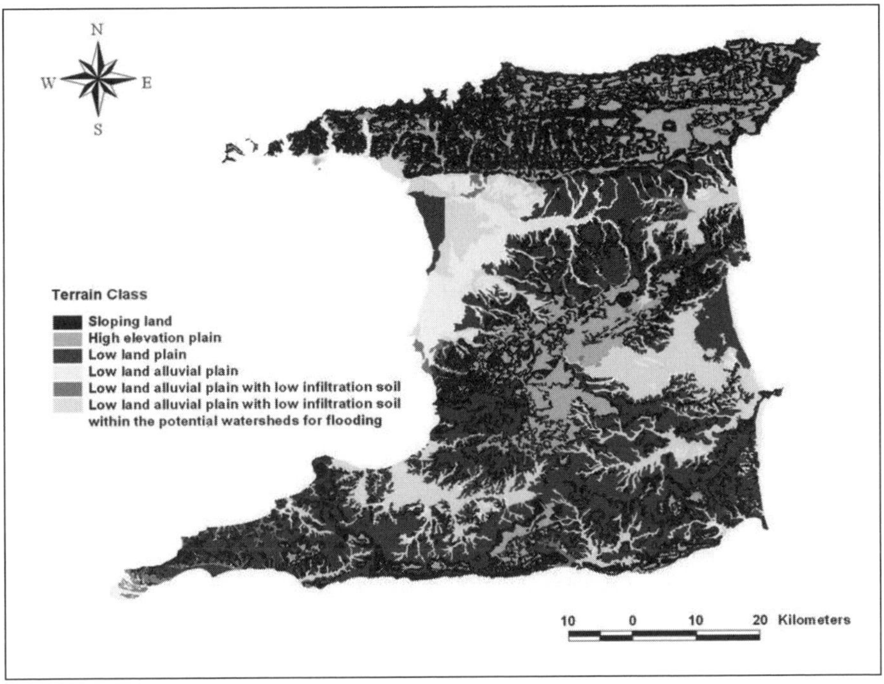

Figure 9.3 Terrain liable for flooding.

liable for flooding, based on terrain characteristics. The sloping lands in the Northern Range, such as the central part and southern border, have very low probability for experiencing flooding, except flash flooding. Contrary to the overall topographic configuration, moderate and high hazard situations are not necessarily located on the higher elevation areas (JBA 2004). Significant areas of Trinidad are plains and are used by residential developments depict moderate to high hazard situation as lowland areas are susceptible to flooding (Smith 1991). Geology and soil maps can provide information on soil series associated with river, lake, wetland and tidal deposition. They can be useful in determining historic flood plains at geologic time scales, but they do not provide any indication of event probability (JBA 2004). Based on the flood event inventory data, lowland alluvial plains show a higher number of events than other geology types, and the alluvial plain with low infiltration soil certainly is more susceptible to flooding for this reason, rather than its other geology and soil characteristics. Most probably, this area within the potential watersheds for flooding, must be the most susceptible area for flooding. It is obvious that the villages and residential areas within these highly susceptible areas have a higher risk of flooding. Specifically, areas with a record of flood events have the maximum hazard risk.

9.4.3 Map of Flood-susceptible Areas

A key element of sustainable flood management requires flood-prone maps to support the proposed risk-based approach for planning and development. This study identified three levels of susceptible areas based on the terrain characteristics, namely, susceptible, moderately susceptible and highly susceptible. The flood events were also categorized into three classes based on the last 20 years of event data as low-, moderate- and high-frequency events.

In Figure 9.4, the north and southwestern areas of Trinidad are located in highly susceptible flood-hazard zones, due to the high-frequency flooding events. This zone includes the two main cities (Port

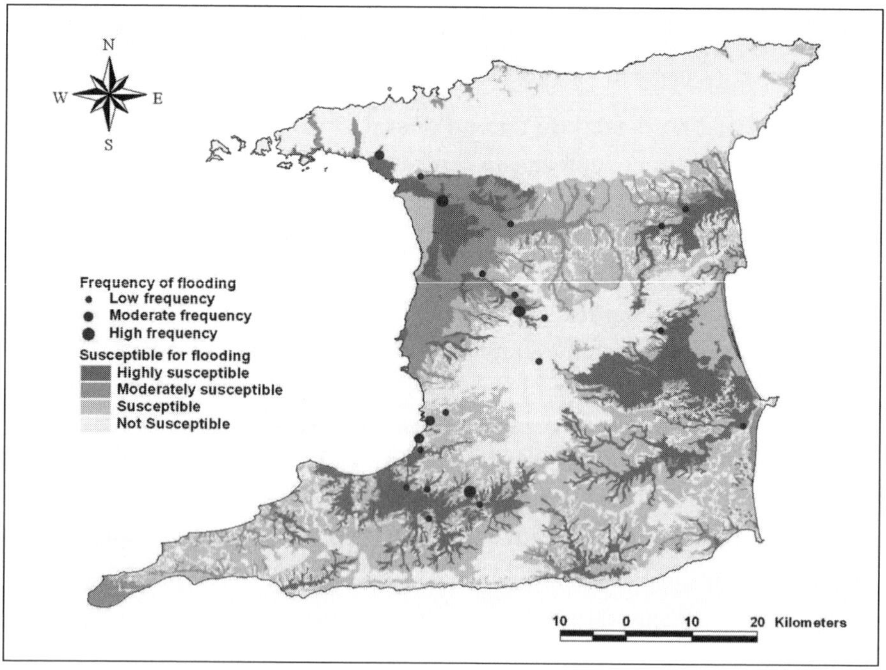

Figure 9.4 Flood-prone map of Trinidad.

of Spain and San Fernando) in Trinidad. Other than these cities, a number of towns and villages, such as Caparo, Caroni, Barrackpore and Marabella, are frequently flooded. Some areas in eastern Trinidad, though they are located in highly susceptible areas seem to have a low frequency of flooding perhaps due to being mostly covered by forest and are therefore comparatively low flood-hazard zones.

9.4.4 Evaluation and Discussion

The map was evaluated using the historical data set with the number of occurrences that may indicate the susceptible area for flooding (Table 9.10). When the locations with three or more flood events were considered, 86% of the locations (19 of 22) were found to be in the region of highly susceptible areas for flooding. Two locations, San Fernando and Marabella, are in the susceptible areas for flooding, but the high frequency of flooding in these areas could be because of pluvial flooding.

A critical flood event shows all the areas that are likely to flood during a major flood. The definition of a critical flood event is somewhat subjective, and its description may vary depending on the purpose of the investigation. In all cases it is defined by four features (Smith 1991; Cooper 2001):

1. Rainfall depth
2. Duration
3. Time-distribution of the rainfall depth over the duration of the storm
4. Spatial distribution or the area over which the features in (1) to (3) remain unchanged

The chosen critical flood event was the flood that occurred on 14 July 2005. This was the major flood that occurred after the devastating hurricane, Emily. The locations (communities, roads and areas) that were flooded during this event were Santa Cruz, Saddle Road Roundabout, Siparia, Marabella, Mamoral, East Dry River, Maraval, Woodbrook Youth Centre, South Quay, Port of Spain, Churchill Roosevelt Highway, Arima, San Raphael, Caroni, Talparo and Caparo.

Flood-event mapping (demarking these areas) requires detailed DEM, satellite imagery during the flooding time and high levels of

Table 9.10 Flood Locations and Their Terrain Susceptibility Level

Location	Flood Occurrence	Terrain Class	Susceptibility Level
Caparo	13	6	Highly susceptible
Caroni	11	6	Highly susceptible
Barrackpore	11	6	Highly susceptible
Saddle Road Port of Spain	10	6	Highly susceptible
Marabella	8	3	Susceptible
San Fernando	8	3	Susceptible
Penal	6	6	Highly susceptible
Sangre Grande	5	6	Highly susceptible
Debe	5	6	Highly susceptible
St Helena	4	4	Moderately susceptible
Biche	3	6	Highly susceptible
Oropouche	3	6	Highly susceptible
Mayaro	3	6	Highly susceptible
Woodland	3	6	Highly susceptible
Rochard Douglas Road	3	6	Highly susceptible
Gasparillo	3	3	Highly susceptible
Mamoral	3	2	Highly susceptible
Palmiste	3	6	Highly susceptible
Todd's Road	3	6	Highly susceptible
Barataria	3	6	Highly susceptible
Tabaquite	3	2	Highly susceptible
Longdenville	3	4	Highly susceptible

ground data such as flood height, rainfall, river flow and so on. A major deficiency of such mapping in Trinidad at this time was the unavailability of high-resolution detailed DEM cloud-free satellite data and during the critical flood event, as well as the difficulties in obtaining the field data in the amount required. Nevertheless, such event mapping should clearly identify flood-prone areas and could have a role to play in validating any new national flood mapping (Canisius et al.

1998). A major drawback to event mapping is the fact that the probability of each outline will be particular to that event and may change through a catchment for the same event (JBA 2004). Second, the number of unexpected variations from previous flood data could be due to pluvial flooding.

Examining the total flood locations indicates that about 30% of the locations are in urban areas, which are very small (6.2%) compared with other land cover types (based on land use/cover classified image). This indicates the flooding that occurs in these areas is pluvial flooding and related to, or characterized by, rainfall. Typically, it can occur in short duration storms and can be associated with sheet or surface flow and flooding in urban areas, when rainfall exceeds sewer capacity or when a route to a watercourse or drainage channel is interrupted (Baban and Kantasingh 2005).

9.5 Conclusions

This study shows a simple and cost-effective methodology, which uses GIS for creating flood-prone maps from the available data on a national scale. It is acknowledged that accuracy of the key information, past records of flooding and geophysical characteristics depends upon the scale that represents them.

It is anticipated that the flood-prone map developed in this study will be employed for a range of uses, which may include insurance guidance, flood-risk assessments, public awareness of various flood-risk zones, generation of flood-warning schemes, emergency evacuation planning, strategic road network planning, protection of key utility assets, and water resources management and land-use planning.

Another use of an understanding of the effects of inundation over time will be to help in the construction or reconstruction of infrastructure. If a detailed analysis shows specific flood duration patterns, then structures can be built in a way that mitigates impacts from future storms. The inundation history, conditions and duration can also be used to make appropriate future planning scenarios.

Due to lack of cloud-free satellite images and detailed DEM, we could not access the large-scale flood event and duration maps for critical flood events. Detailed study requires a large amount of local topographical and/or detailed data such as LIDAR and weather independent SAR data. This option is likely to be resource intensive to produce national-level maps. However, in some locations, it may be justified. Other than that, further research is required into mapping pluvial flooding at the scales necessary for flood risk mapping purposes.

References

Baban, S.M.J., and J. Flannagan. 1998. Developing and implementing GIS-assisted constraints criteria for planning landfill sites in the UK. *Planning Practice and Research* 13, no. 2:139–51.

Baban, S.M.J., and R. Kantasingh. 2005. Mapping floods in the St Joseph Watershed, Trinidad, using GIS. *International Association of Hydrological Sciences (IAHS)*, publication no. 295:254–64.

Billa, L., S. Mansor, and A.R. Mahmud. 2004. Spatial information technology in flood early warning systems: An overview of theory, application and latest developments. *Malaysia Disaster Prevention and Management* 13, no. 5:356–63.

Bonham-Carter, G.F. 1994. *Geographic information systems for geoscientists.* Kidlington: Pergamon.

Buckley, D.J.A., and W.G. Hendrix. 1986. *Use of geographical information systems in assessment of site suitability for land application of waste. Geographic information systems in government,* vol. 2. Hampton, VI: Deepak.

Cooke, R.U., and J.C. Doornkamp. 1974. *Geomorphology in environmental management: An introduction.* Oxford: Clarendon.

Canisius, F.X.J., H. Kiyoshi, M.K. Hazarika and L. Samarakoon. 1998. Flood monitoring in the central plain of Thailand using NOAA/AVHRR and JERS-1 SAR data. Paper presented at the Twenty- fourth Annual Conference and Exhibition of the Remote Sensing Society. United Kingdom.

Chan, N.W. 1997. Increasing flood risk in Malaysia: Causes and solutions. *Disaster Prevention and Management* 6, no. 2:72–86.

Cooper, V. 2001. *Inland flood hazard mapping for Antigua and Barbuda: Summary report.* St Augustine, Trinidad: Department of Civil Engineering, University of the West Indies.

Environmental Management Authority (EMA). 1996. *State of the environment report: Trinidad and Tobago.* Port of Spain: Government Printery.

Few, R., M. Ahern, F. Matthies and S. Kovats. 2004. *Floods, health and climate change: A strategic review.* Tyndall Centre Working Paper no. 63.

Frasier, J. 2005. Save river and spare us flooding. *Trinidad Express.* 23 November.

Gyulavary, P., and A. Lopez. 1991. Building a geographic information system (GIS) for siting New York City sludge management facilities. In *Civil engineering applications of remote sensing and geographic information systems.* New York: American Society of Civil Engineers Publications.

Heywood, I. 1991. *Managing our future environment: Some GIS issues.* Association for Geographical Information Yearbook. London: Taylor and Francis.

James, M.D. 2004. Be realistic about floods. *Trinidad Express.* 12 September.

JBA. 2004. Requirements for flood mapping: Scoping study. Environment Group research report. JBA Consulting, Scottish Executive.

Jensen, J.R. 1986. Solid and hazardous waste disposal site selection using digital geographic information systems techniques. *Science of the Total Environment* 56:265–76.

Jones, J.L., T.L. Haluska, A.K. Williamson, L. Martha and M.L. Erwin. 1998. Updating flood inundation maps efficiently: Building on existing hydraulic information and modern elevation data with a GIS. US Geological Survey Open-File Report 98-200. http://wa.water.usgs.gov/pubs/ofr/floodgis.

Jones, J.L. 2004. Mapping a flood . . . before it happens. US Geological Survey Fact Sheet 2004–3060. http://pubs.usgs.gov/fs/2004/3060/.

Kenny, J., P. Correau and L. Katwaru. 1997. Extensive extracts from a survey of the biological diversity Trinidad and Tobago. Environmental Management Authority of Trinidad and Tobago. http://www.ema.co.tt/nbsapweb/Kenny-et-al.htm.

Khan, F. 2005. Address to the nation by minister of works and transport. http//:www.ttgov.tt/speehes/speech.asp.

Lyle, J., and F.P. Stutz. 1983. Computerised land use suitability mapping. *Cartographic Journal* 20, no. 1:39–49.

Ramroop, S. 2005. Flooding analysis strategy using GIS. Paper presented at the 2005 ESRI International User Conference. San Diego.

Smith, K. 1991. *Environmental hazards: Assessing risk and reducing disaster.* London: Routledge.

Sharma, V., and T. Priya. 2001. Development strategies for flood prone areas: Case study, Patna, India. *Disaster Prevention and Management* 10, no. 2:101–9.

Ward, R. 1978. *Floods: A geographical perspective*. London: Macmillan.

Wohl, E.E. 2000. *Inland flood hazards: Human, riparian, and aquatic communities*. Cambridge: Cambridge University Press.

SECTION 3

Geohazards Management

CHAPTER 10

Developing a Proactive Approach to Geohazards Management in Trinidad and Tobago

SERWAN M.J. BABAN

Abstract

The economic, environmental and social costs of annual flood and landslide events amount to millions of dollars in Trinidad and Tobago. Additionally, in November 2004, floods and landslides, for the first time, caused loss of life in Tobago. The current management of floods and landslides is reactive since the major effort is focused on post-event cleaning-up operations. Mitigation works are designed to repair infrastructure after the event has occurred. Clearly, there is an urgent need for objective decision making, and for moving geohazards management from being reactive to proactive. However, the lack of an effective and reliable information base makes this transformation difficult. For example, at present there is an absence of a national data depository for hazard events in which event occurrences can be recorded and quantified for post analysis. Furthermore, several governmental agencies seem to be responsible for geohazard management. These agencies are neither capable of handling geohazards on their own nor do they have effective coordination among themselves.

This chapter advances the concept for a proactive approach in Trinidad and Tobago through developing the use of a holistic approach which utilizes available and reliable cutting edge technology such as geoinformatics (satellite remote sensing, global positioning systems and geographic information systems). Furthermore, careful attention needs to be directed towards the development of a national emergency strategy; establishing national geohazard inventories and databases; developing early warning systems; using predictive understanding of the processes and triggering mechanisms of landslides and floods; and setting international standards for all consultancies and research projects funded through the government ministries and agencies. In addition, there is a need to develop effective programmes for public awareness, education and information, as well as enhancing the implementation capabilities of relevant government agencies.

10.1 Introduction

Every year, geohazards (floods and landslides) account for economic losses of approximately tens if not hundreds of millions of dollars in the Caribbean region including Trinidad and Tobago. Additional costs in terms of disruptions to the social fabric, the damage to the flow of goods and services (for example, lower output from damaged factories, lost productivity and the like), and short- and long-term impacts on the environment and economy remain unquantified.

The states in the Caribbean have a number of common characteristics, which make them vulnerable to geohazards. These include geography climate/weather conditions; limited physical size; finite natural resources; dependence on agriculture and tourism; and high population densities concentrated in vulnerable areas, including hillsides and flood plains. In addition, the region is experiencing rapid economic development combined with a fast rate of urbanization, population growth and questionable agricultural practices. Typically, these factors lead to floods, landslides, deforestation, soil erosion and extinction of an unknown number of animal and plant species (Baban 2003). Clearly, there is a need to determine strategies to cope with the uncertainties of the effects of geohazards on natural resources, the environment, and goods and services (Baban et al. 2004). This process will require an understanding of the links between geohazards and societal well-being, and the promotion of effective interventions (Baban 2001, 2004).

Reliable information is one of the most important strategic factors influencing decision making and development. However, the Caribbean region, like other developing regions, suffers from a scarcity of reliable and compatible data sets. For example, the soils and the land use/cover maps for the region were constructed during the 1950s and 1960s. Furthermore, in the case of Trinidad and Tobago, the national hazard response is expressed in a series of Hazard and Response Maps for the islands at scales varying from 1:150,000 to 1:75,000 (Baban 2004). This map series established nine hydrometric regions for Trinidad and one for Tobago. It provides the geographic locations of areas likely to be affected by floods, as well as the locations of landslides and environmental hazards. This map was developed based on the perception of "accepted areas" prone to hazard events, and it was never verified nor quantified. In fact, collecting, analysing and verifying data in the field has proven to be very difficult.

Nevertheless, there are clear indications that the information poverty obstacle can be overcome by using reputable technologies that facilitate management decisions, such as geoinformatics, which encompass remote sensing, geographic information systems (GIS) and global positioning systems (GPS). Geoinformatics contains the necessary tools to collect, handle and analyse the necessary data sets, as well as to expand our knowledge of the processes involved at the appropriate scales (Baban et al. 2004). Geoinformatics can provide a means for the unconstrained analysis of conditions that influence landslides and floods. Furthermore, spatial and temporal distribution patterns not apparent within written reports can be highlighted through appropriate maps, and early warning systems can be developed, which would help to provide effective information for geohazards management.

However, it should be indicated that the insufficient funds allocated to obtain information in developing countries is a serious obstacle. In fact one of the barriers to sustainable development in developing countries is lack of information requisite to planning it (Bernhardsen 1992). Evidently, geohazards can raise awareness of risk and the need for management. Sadly, in recent times, Trinidad and Tobago has suffered a series of high-intensity rainfall events. These resulted in severe flooding and landslides, which caused an unprecedented loss of lives during November 2004. This fact must be used as a wake-up call to the seriousness of the situation to create pressures for increasing expenditure on geohazards management. In other parts of the world, for instance, in the United Kingdom, the London and East Coast floods of 1953 created public support for investing £750 million in the Thames Barrier to control future floods affecting the capital city (Twigg 2002). It should also be articulated that geohazard management is cost-effective and can lead to significant reduction in destruction and loss during geohazard events (Table 10.1).

The aim of this chapter is to indicate that a real option for transforming geohazard management in Trinidad and Tobago from reactive to a proactive mode can be through developing a holistic approach. In addition, this approach can form the basis for developing a national emergency strategy and short- and long-term priorities, including the establishment of national geohazard inventories and databases; the development of early warning systems; the development of effective programmes for public awareness, education and information; and the enhancement of the implementation capabilities of relevant government agencies. Finally, this strategy will need to implemented by utilizing reli-

Table 10.1 Cost-effectiveness of Mitigation and Preparedness

- The World Bank and US Geological Survey calculated that economic losses worldwide from natural disasters during the 1990s could be reduced by US$280 billion, if US$40 billion were invested in mitigation and preparedness.
- The Thames Barrier in London cost £750 million, but the potential loss of property to floods without the barrier was estimated to be £3.5 billion.
- In Vietnam, 12,000 ha of mangroves planted by the Red Cross protect 110 km of sea dykes. Planting and protection cost US$1.1 million but has reduced the cost of dyke maintenance by US$7.3 million per year (and the mangroves have protected 7,750 families living behind the dyke).
- According to Oxfam, the value of cattle saved on a flood shelter of approximately 4 acres in Bangladesh during the 1998 floods was as great as Tk 4,000,000 (£150,000) against a construction cost of only Tk 700,000 (£8,560).

Source: Twigg 2002.

able cutting edge technology such as geoinformatics to overcome information poverty.

10.2 Challenges Facing Geohazards Management

Managing geohazards in developing countries presents a number of challenges, including the following:

1. The fact that the focus is often on campaigns to satisfy short-term necessity. The unfortunate result is that overviews of natural resources and environmental status are lacking. The information requisite to planning and prudent management of renewable and other resources is inadequate for decision making that provides effective and timely answers (Bernhardsen 1992; Baban 2003).
2. Collecting data on the real costs of geohazards is inherently problematical. Conventional monitoring and evaluation methods are primed to understand and measure something that has taken place as the result of an intervention (Twigg 2002). However, in the case of geohazards, what needs to be done is to estimate the cost in cases where a successful evasion or a reduced impact results as a consequence of preparedness and mitigation measures.
3. Geohazard management professionals tend to calculate the consequences of particular actions with a degree of objectivity, but perceptions of risk are not entirely rational; they are shaped by numerous social, cultural and psychological factors, which cannot be grasped fully. Among the factors that can form perceptions are whether the risk can be controlled; whether the exposure is voluntary; if the risk is familiar, the potential for catastrophes; whether the consequences are greatly feared; and who is at risk, or could benefit from risk reduction measures.
4. In the majority of cases, data are collected, compiled and handled for a specific project with specific purposes with no regard to other uses. Furthermore, the qualities of these data collections and the acquisition methods employed are frequently unknown. As a result, the data available are questionable and unmanageable and thus, cannot be used productively. Besides, monitoring and evaluation of

mitigation and preparedness tend to be tied to the project cycle. As a result, they tend to focus on outputs rather than impact. Furthermore, post-project studies of long-term impact are rare (Twigg 2002).

5. The community of people working in geohazard management is fragmented along disciplinary and institutional boundaries (Table 10.2). The coordination tends to be poor between these communities. This situation clearly presents an obstacle to formalizing a deeper understanding of costs and benefits, as well as the concepts and methods concerning geohazard management.

Table 10.2 Disciplinary and Institutional Groups Involved in Risk Reduction

Disciplinary Groups
• Physical scientists (geologists, hydrologists, meteorologists, etc.) • Social scientists (psychologists, sociologists, economists, etc.) • Engineers • Architects • Doctors • Nutritionists • Planners • Humanitarian relief workers
Institutional Groups
• Multilateral and bilateral aid agencies • National government agencies • Local government agencies • Civil society organizations (NGOs community-based organizations, trades unions, etc.) • Academic and educational institutions • Private sector (insurance, engineering, IT, consultancy, etc.) • Media

6. The misunderstanding and mismatch of priorities between researchers and programme managers on timing actions and on how much knowledge is needed before action can be taken. The former, typically more detached and analytical, want to understand a situation as fully as possible before intervening; the latter, typically much closer to communities, are keen to get involved with the minimum of delay.
7. Decisions about investment in geohazard management are essentially political ones. Policymakers evaluate not only the practicality of suggested management options but also their desirability: desirability is basically the political element in the equation. Thus far, central decision makers have not emphasized the need for environmental data. This lack of political interest coupled with inadequate overviews of natural resources results in uninformed decisions being made. This is the case whether governments, international aid agencies or NGOs make them. Research and analysis of problems and solutions are very important influences, especially in democratic societies, but their influence should not be overstated.
8. Global, regional and national financial aid organizations have only been moderately helpful, as their involvements have usually been limited to short-term projects of limited geographic extent. Furthermore, data collection and validation, as well as analysis, are in the format of reports and seldom arise to the level of research.

10.3 Developing a Proactive Approach to Geohazards Management in Trinidad and Tobago

The increasing level of geohazard events in Trinidad and Tobago accelerates the need for up-to-date information for geohazard management. The current response to floods and landslides can simply be categorized as reactive since the major effort remains in post-event cleaning-up operations . Mitigation works are designed to repair infrastructure after the event has occurred and there is a heavy reliance on a "hard engineering" approach.

Information poverty, with respect to landslides and floods, has restricted the level of response of the state towards hazard reduction and geohazard response. At present, there is an absence of a national data depository for hazard events, where event occurrences can be recorded and quantified for post analysis. Furthermore, the fragmented governmental agencies, under whose responsibility hazard response and reduction falls, are similarly deficient in their recording of hazard events. The approach of assuming hazard events as "Acts of God" are long past, given the global developments in hazard management, planning, mitigation and risk analysis. Careful attention is required in construction planning in order to minimize the potential risk of damage to life and property. The limited resource base, coupled with demands to prioritize responses to natural hazards, necessitates the development of a national agenda that can provide both short- and long-term solutions for the benefit of the people of Trinidad and Tobago.

A good national agenda should be focused on a holistic approach and comprised of a range of diverse activities: organizational, educational, structural, economic and so on. These activities need to be mutually reinforcing; for example, training in safe building techniques should be complemented by regulation of land use and both the setting and enforcement of building standards, as well as by measures to address the economic and social pressures that force poor people to live in pitiable housing in hazardous locations. As a consequence, there is a need to determine strategies for geohazard management and to cope with the uncertainties of the effects of geohazards on the environment, economy and society at large. This process will require the following:

1. Developing a scientific understanding of the processes involved and establishing a holistic approach for geohazard management that can provide answers for key questions:
    - How will geohazards affect the environment, economy and society at large?
    - How vulnerable are these sectors in different parts of the country?
    - Which different community groups will be most at risk?
    - How might different parts of the country adapt to cope with geohazards?
2. Establishing a priority list for requirements and deliverables including the following:

- developing early warning systems for landslides and floods and using predictive understanding of landslides and floods processes and triggering mechanisms;
- setting international-level standards for all consultancies and research projects funded through government ministries and agencies;
- making data validation and metadata development an essential part of all the outcomes from consultancies and research projects funded through the government ministries and agencies;
- coordinating the efforts among various relevant ministries and agencies concerned with landslides and floods;
- developing effective programmes for public awareness, education and information dissemination;
- adopting a public education drive in schools and through the media to inform the public of "incorrect" practices leading to geohazards and the consequences for the community and the country; and
- examining and developing effective insurance mechanisms as an instrument for managing geohazards and employing loss reduction measures.

3. Identifying the tools for collecting and managing the necessary national geohazard inventories and databases. The lack of a reliable inventory of events is a primary setback to the evolution of a proactive response to hazard mitigation. The development of new maps to provide a realistic reflection of the on-the-ground situation requires a systematic approach, which will avoid current pitfalls. This may be accomplished through the establishment of a central agency responsible for a single repository for information regarding hazard events. As an interim measure, the mapping of events should adopt geoinformatics technologies that are scientific and economically feasible. The independently piecemeal development of data sets and applications without metadata standards results in duplication of efforts and wastage of scarce financial resources from agencies purchasing the same data repetitively. The availability of recent high-resolution imagery and topographic data sets, from which national inventories are derived, should be pursued. In the case of

floods, the acquisition of remotely sensed data soon after events will provide an effective means of quantification the extent of the event.

4. Developing plausible management scenarios to account for the effects of geohazards on the environment, economy and the society at large.
5. Developing a regional agenda to deal with geohazard management with a sharp focus on small island estates issues such as recovery and resilience. Furthermore, attention should be given toward assisting the most vulnerable and poorer sections of society so that they can build up livelihoods that are more sustainable and less vulnerable to external forces by accumulating and using a variety of "assets": human, social, physical, financial and natural capital.
6. Attempting to persuade international funding agencies to fund quality research and capacity building, in a meaningful and sustainable fashion, to assist Trinidad and Tobago specifically with the proactive decision making process in managing geohazards.

10.4 Conclusions

Trinidad and Tobago is experiencing a rapid rate of population growth, economic development and urbanization, which could potentially lead to aggravated geohazards. All of these socioeconomic factors are placing increasing pressures on the environment, economy and development of society. Consequently, there is a real need to understand the impacts of geohazards with the view to design intervention strategies to manage it. This objective can be realized through establishing a reactive and a holistic management approach driven by a single agency to coordinate, manage and deliver a specific set of priorities including quality data collection via geoinformatics, which contains the necessary tools (remote sensing, GIS and GPS) to collect, handle and analyse the necessary data sets, as well as through developing early warning systems. Additionally, the focus must be on assisting the most vulnerable groups in society and to develop practical analytical methods, that is, it must be possible for a wide range of geohazard professionals and institutions to use them, and these groups must see an advantage in doing so. Finally, it is acknowledged that improved data and analysis are not enough by themselves. There should be a vigorous effort to use results together with other campaigning methods in order to influence decision makers at every level.

References

Baban, S.M.J. 2001. Managing the environment in the Caribbean region using remotely sensed data and GIS. *Proceedings of the Urban and Regional Information Systems Association (URISA) Caribbean GIS Conference*, 202–13. Montego Bay, Jamaica.

———. 2003. Responding to the effects of climate change on agriculture, fisheries and tourism in the Caribbean region utilising geoinformatics. *Journal of Farm and Business* 5, no. 1:95–111.

———. 2004. Attaining a balance between environmental protection and sustainable development in the Caribbean region using geoinformatics. *West Indian Journal of Engineering* 26, no. 2:22–34.

Baban S.M.J., B. Ramlal and R. Al-Tahir. 2004. Issues in information poverty and decision-making in the Caribbean region: A way forward. *West Indian Journal of Engineering* 27, no. 1:28–37.

Bernhardsen, T. 1992. *Geographical information systems*. Arendal, Norway: Viakt IT.

Twigg, J. 2002. Lessons from disaster preparedness. Paper presented to the International Conference on Climate Change and Disaster Preparedness. The Hague, Netherlands.

CHAPTER 11

Issues in Flood Risk Management

ANDREW FOX

Abstract

Flood events are a significant and increasing threat to populations in the Caribbean. As a geohazard they are not new, but they do present many unique and complex challenges for governments and the experts that advise governments. In order to inform those with an interest and desire to see geohazards better managed, this chapter sets out to explore these challenges and draw lessons from international research and experience with the aim of proposing strategies that could aid the development of more resilient and robust management systems.

11.1 Introduction

Research has shown that 13 of the top 25 countries to have suffered from the impact of natural disasters were Small Island Developing States (SIDS) (Briguglio 1995; Pelling and Uitto 2002). These SIDS are generally grouped into four main clusters:

1. Caribbean Island States
2. Pacific Island States

3. Indian Ocean Island States
4. West African Island States

Pelling and Uitto (2002) identified that the relative prosperity of small islands is often built upon fragile economic and political foundations, making them especially vulnerable to geohazards. This vulnerability has been recognized for a number of years. According to Briguglio (1995), the problems faced by small islands were first raised to the attention of the United Nations in 1972 during UNCTAD III, and by 1988 a wide array of disadvantages were recognized. He went on to analyse the factors that render these economies vulnerable, namely small size, insularity, remoteness, proneness to natural disasters, highly limited internal markets and dependence on foreign sources of finance. From this, Briguglio developed a global index, which he used to rank states in order of their vulnerability to disasters. In doing so, he determined that nine out of the ten most vulnerable states to geohazards were small islands. The top ten states most vulnerable to geohazards were Antigua and Barbuda, Tonga, Seychelles, Vanuatu, St Kitts and Nevis, St Lucia, Chad, Singapore, St Vincent, and Grenada.

As a result, the vulnerability of these Caribbean states is becoming more widely recognized. In a paper looking at climate sensitivities and the adaptation measures needed to improve resilience in Jamaica's urban water management sector, Jones and Spence (2003) described how this high vulnerability, coupled with low adaptation strategies in sectors such as water resources, tourism, agriculture and fisheries, is highlighting the need for an integrated approach to geohazard risk management. In small islands, "trends such as rising seawater levels, increasingly variable rainfall, accelerating storm water runoff, increasing demand for water, and increasing pollution of surface, ground and sea water are so potentially disturbing that they threaten the development of their economies and the health of their people" (Third World Water Forum, http://www.adb.org/Water/Theme/small_islands.asp).

This view was similarly expressed by the main disaster management agency in the Caribbean region, Caribbean Disaster Emergency Response Agency (CDERA), at the Second World Conference on Disaster Reduction in Kobe, Japan, in 2005. "CARICOM recognizes the inextricable link between poverty and disasters and accepts that poverty alleviation and sustainable development cannot gain traction without a strengthened commitment to disaster reduction at the regional, national, community and individual levels in each member country" (CDERA 2005, 26). In particular, in 2005, CDERA outlined how the Caribbean region suffers repeated and extensive losses from hurricanes and their associated wind, rain and storm surge damage. They put forward a programme to address this vulnerability, which focused on improving the resilience of nations and communities to hazard impacts. Resilience was seen as being advanced through the following themes: governance (institutional and policy frameworks for risk reduction), knowledge management, community disaster planning, flood management and adaptation to climate change.

11.2 Floods and Flooding

Of the range of geohazards occurring in the Caribbean, floods are the most serious in terms of frequency of occurrence and the extent of damage caused (CDERA 2005). Between 1996 and 2000, 90% of all the Caribbean states were affected by floods but only 25% had flood hazard management plans. In 2003, flooding accounted for 60% of the total economic losses from all types of disasters in the Caribbean region (CDERA 2005), and as a result of this knowledge, a list of priority areas for dealing with flood risk was developed: Identify flood hazard areas, map flood hazard, assess vulnerability and establish early warning systems (CDERA 2005).

On the global scale it has been estimated that one third of the total economic losses that occur are as a result of flooding. Of all the natural catastrophes worldwide, two thirds of the world's population are affected by flooding. In fact according to Pilon (2002) and a study by the World Meteorological Organisation (WMO), during the 1990s, the total number people across the world who had been affected by floods reached a total of 1.5 billion. This trend has continued into the new millennium with the year 2002 highlighted as a particularly bad year for flooding; there were millions of people affected in Europe alone (WMO 2004).

Floods do not just cause property damage, they can also result in psychological trauma for the victims, damage to the environment from pollution and increases in water-related disease. They can also lead to development blight if the insurance industry refuses to provide coverage to regions subject to ever increasing flood risk (EC 2004). Floods can last minutes (flash floods) or weeks and months in larger water systems. Flooding is not a new phenomenon, as the European Environmental Agency defines them: "Floods are natural features of running water systems" (EEA 2001). In this definition, a water system is deemed to include the water channel, as well as the flood plains surrounding the channel. Other authors have been more explicit, explaining how natural water channels only have a limited capacity for storage, so flood plains are an essential feature of the water system, providing the temporary storage required as the flood wave passes (Wilson 1983; NRA 1994). Loss of the storage capacity on the flood plain often leads to problems elsewhere in the water system. For this reason, to truly understand the nature of floods, the above definition lacks some level of detail, and a second definition from the same source is provided: "extreme water runoff during which human lives, property, infrastructure and economy are threatened" (EEA 2001). This is a more useful definition because it helps to identify an essential feature of floods that needs to be considered before a proper study can be done. This feature is that flooding only becomes a concern when it interacts with the human social environment.

The focus on water runoff in the above definition is slightly misleading since, according to the WMO (2004) and Lancaster et al. (2004), floods are caused by a variety of phenomena: rain, snow melt, tides, storm surges, wave action, dam breaks, rising ground water, overland runoff, drainage or infrastructure failure.

Depending on where they occur, floods may be classified as:

1. Fluvial (river flooding)
2. Coastal
3. Estuarial, or
4. Pluvial (overland flooding)

Nicholls et al. (1999) examined the potential impact of global sea level rise on coastal flooding. They modelled a 1 m rise in sea level and determined that such an event would threaten 50% of the global coastal wetlands and also result in a trebling of the population at risk from storm surges. In this case, coastal wetlands were defined as salt marshes, mangroves and inter-tidal areas. The loss of coastal wetlands due to sea level rise can be offset by inland wetland migration, but in areas without low-lying coastal areas, or where human development prevents wetland migration, coastal squeeze occurs.

The research predicted that areas with low tidal ranges are more vulnerable than areas with high tidal ranges, and this particular risk has significant relevance to many of the small islands states (Nicholls et al. 1999). Large increases in the number of people affected by flooding will occur in these small island states, and they were predicted to experience the largest relative increase in flood risk of any region. By 2080, the average annual number of people experiencing coastal flooding in small islands was predicted to increase by more than 200 times the numbers in 1999 (Nicholls et al. 1999).

As mentioned above, human action can do much to exacerbate or mitigate the incidence of flooding, and much research has been undertaken to investigate the links between human actions and the incidence of flooding (EEA 2001; EC 2004; Lancaster et al. 2004). From this research the main actions that have been found to have a direct link to the incidence of flooding are as follows:

1. Flood plain development
2. River channel alteration
3. Deforestation
4. Land drainage
5. Improper design of flood protection schemes
6. Industrial development and pollution

Hamill (2001) introduced some other issues that have recently raised themselves for concern and may ultimately be added to the above list. First, there are the actions of governments in the field of cloud seeding, which aims to promote rainfall in areas prone to extensive dry spells. This has led to court actions by members of the public in the United States. Second, irrigation in association with land drainage has the effect of raising ground water levels and, in some cases, polluting soils, so killing vegetation (salinization). Finally, Hamill points out that acid rain (associated with industrial pollution) can drastically alter the landscape by killing both aquatic life and vegetation, so affecting the incidence of flooding.

11.3 Precipitation, Runoff and Hydrology

The Caribbean rainfall season occurs mainly from May to November, and most territories have distinct wet and dry seasons (Jones and Spence 2003). Laing (2004) describes one of the greatest problems facing meteorological forecasters in this region – the prediction of heavy rainfall and flash floods: "Forecasting in the tropics presents significant challenges and especially on islands where rainfall can be intense and highly localised". Laing goes on to say that much attention is sometimes given to the destruction caused by hurricanes, but these events are relatively rare for any single country. Far more important is the damage caused by smaller and more frequent "heavy" precipitation events that also cause death and serious economic damage but for which little international assistance is available.

In terms of rainfall, there are two broad classifications with relevance to this chapter. *Orographic* precipitation results when moist air from the sea meets the land mass. This precipitation is associated with frequent low-intensity storms. *Convective* precipitation occurs when warm air rises over the land mass, forms clouds and then condenses. Thunderstorms are generally convective and are often associated with heavy rainfall events and flash flooding (Hamill 2001).

The steep topography of Caribbean islands often aids heavy rainfall through orographic lift, or by creating persistent low-level convergence zones that act as sources for new convection, resulting in precipitation (Laing 2004). In general, large amounts of rainfall occur over the first area of high ground encountered with an almost linear increase in rainfall with increasing wind speed. With few exceptions, the prediction of heavy rainfall over Caribbean islands has been handicapped by the lack of a high-resolution observation network needed to provide guidance on small spatial scales. This has resulted in the development of a number of statistical methods of analysis that combine satellite imagery with infra-red and micro-wave sensors to detect heavy rainfall. But these techniques require frequent and reliable access to technology, which is not always possible in small island states (Laing 2004).

When predicting rainfall runoff, Wilson (1983) identified that a range of environmental factors, in addition to the topographic features identified above, can affect the amount of runoff. These include the following:

1. Geology
2. Temperature
3. Humidity
4. Wind
5. Radiation
6. Soil moisture deficit

A catchment's features, such as its total area and shape, are also important in determining volumes of runoff, and the frequency of streams and rivers can affect the magnitude and duration of peak flows (Wilson 1983).

In relation to the Caribbean, Jones and Spence (2003) described the inner arc of volcanic islands as having a steep conical-shaped topography and generally radial drainage patterns with steep, well-incised channels that carry substantial runoff during rain events. They described the outer arc of islands as generally having a more low-lying topography with minor gradients and a greater propensity to ponding on the surface.

For accurate flood prediction, rainstorms themselves also require careful analysis. Duration and intensity variations (resulting in *hyetographs*), coupled with prolonged observation of water channel flows (resulting in *hydrographs*), allow a correlation to be developed between any particular storm profile and a flood occurrence. The study of rainfall in this context is termed hydrology and can be defined as "the study of the occurrence and movement of water on and over the surface of the earth" (Wilson 1983). Hydrologists play a central role in the prediction of floods, their frequencies, durations and intensities. In this chapter, it is not the intention to explore in detail the science and practice of hydrological forecasting but to outline the elements of that discipline, so it is possible to enable an understanding of how and where it fits into the overall flood risk management framework. The work of the hydrologists was outlined in some detail by Wilson (1983), and he was able to develop a description of the range of practices as well as allied academic subject areas that feed into the discipline. Allied academic subjects in hydrology include physics, meteorology, oceanography, geography, geology, hydraulics, statistics, economics, agriculture, forestry, botany and biology.

This wide array of knowledge is applied, in the case of flooding, to the practice of "flood routing", which is the measure of how a flood wave travels along a water channel. The hydrologist calculates storage capacity in the water channels and flood plains for different levels of flood and predicts the probability and magnitude of "n-yr" events (Wilson 1983; NRA 1994; EEA 2001). The n-yr event is an important parameter in flood forecasting, and it refers to the probability that a stated flood level will occur within "n" years and is often referred to as the "return period". Structures are often built to withstand a flood of a certain return period within their design life. The n-yr return periods is often used to control development by zoning areas subject to a particular level of flood risk (NRA 1999):

Zone A – High-density urban areas, return period 1 in 100 for non-tidal and 1 in 200 for tidal

Zone B – Medium-density urban areas and some agriculture, 1 in 75 for non-tidal and 1 in 150 for tidal

Zone C – Low-density urban areas and rural communities with high production agricultural land, 1 in 25 for non-tidal and 1 in 50 for tidal

Zone D – Arable farming, isolated properties and medium production agriculture, 1 in 10 for non-tidal and 1 in 20 for tidal

Zone E – Grassland with few properties and low-production agriculture, 1 in 1 for non-tidal and 1 in 5 for tidal

The study by Jones and Spence (2003) into the resilience of Jamaica's urban water management sector identified a number of problems associated with runoff from heavy rainfall events. These include turbidity levels in excess of handling capacity of water supply installations; pollution of water sources occasioned by the washout of pit latrines; flood water infiltration of well heads; flooding of pump houses; blockage of water supply intakes; washout of pipelines; and reduced storage capacity in reservoirs and tanks due to heavy sedimentation. Difficulties in dealing with these problems were exacerbated by the way in which water resource management is structured in the Caribbean, often characterized by a multiplicity of organizations operating independently and without any clear mechanism for integration. In addition, they identified that there is a lack of data and technical capability from which to develop an effective policy. Policy development that does exist suffers by the lack of a participatory approach (Jones and Spence 2003).

11.4 Flood Planning and Management

Strategies for dealing with flooding often make a distinction between structural or "hard" options and non-structural or "soft" options (MAFF/WO 1993; EEA 2001). Structural flood protection measures include embankments, pumping stations, controlled storage, reforestation, channelization, riparian zoning and hydraulic structures. The problem with these measures is that they require a large capital outlay, often provide a false sense of security and can encourage further inappropriate development. Another problem is that responsibility for flood defence planning is often dispersed between several agencies, thus emphasizing the need for strong inter-organizational coordination and other non-structural measures (WMO 2004).

The focus for non-structural measures is on vulnerability reduction and, as such, these measures should be practical, appropriate and sustainable for the community they serve (Pilon 2002). Non-structural flood protection measures – include land-use control, building development control, forecasting, warning and awareness raising, response planning, and integrated water resource management (IWRM) systems.

Foremost among the non-structural measures is IWRM, which was conceived following the 1992 Dublin and Rio de Janeiro conferences on the environment (TAC 2000). The conferences identified problems in water governance, and, in particular, its fragmented nature that has led to uncoordinated development. Water resources were seen to be managed by top-down institutions, whose legitimacy and effectiveness was increasingly being questioned, and the emphasis in planning was biased towards development with scant regard to protecting the environment (TAC 2000).

IWRM attempts to provide a more balanced view of flood planning, by giving higher recognition to the interrelated and often detrimental effects that development in one area of a catchment may have on others elsewhere in the same catchment (EC 2004; Lancaster et al. 2004). Those in favour of IWRM point to lessons learned from recent flood events, which are focusing not just on the damage caused to individuals and homes, but also on the loss of trade suffered by businesses, as well as the disruption caused to transportation, infrastructure and other essential services (EA 2001).

Furthermore, new technologies such as geographic information systems (GIS) and remote sensing are increasingly being used to manage floods through examining and analysing the spatial and temporal patterns of floods and developing management scenarios for specific catchments, based on spatially variable geomorphologic parameters such as soil characteristics and land-use change, vegetation cover and type, rainfall intensity, and rainfall frequency (Shih 1996; Melesse and Shih 2000). Moreover, the GIS can allow a manager to handle and analyse the spatial data sets, to determine which factors have what effect on floods and to foresee the resulting consequences (Baban and Kantasingh 2005). In terms of the impacts of land use/cover on flooding, GIS can not only be used to map change detection, but also be used to identify trends, both visually and statistically, between land-use changes and flooded areas (Mamat and Mansor 1999). The use of a GIS also allows further spatial analysis of the data derived from remotely sensed images and improves the analysis of the impact of land cover change on regional sustainable development (Xiuwan 2002). In a recent effort, the Government of Trinidad and Tobago has developed scientifically based geoinformatics-driven (GIS and remote sensing) criteria for determining suitability of hillside areas to accommodate built development within hydrological catchments. This is intended to facilitate the development of a national planning policy for hillside development (Baban et al. 2007).

Beyond this, flood risk managers are beginning to recognize the importance of the widely predicted expectation that flooding will increase in magnitude and frequency as the effects of global climate change take effect.

The new era of flood risk management is seeking to achieve a better balance with nature. This includes highlighting the beneficial effects floods have on the environment (such as improving biodiversity) and enhancing these in any management strategy, often by aiming to replenishing wetlands and fisheries (EA 2003; WMO 2004). There is a widespread acknowledgement that flood risk planning cannot eliminate the risk of flooding altogether; rather, it should aim to reduce and maintain risks at an acceptable level (EA 2003). This is an approach that requires a balance between structural measures and non-structural measures and promotes the concept of multi-hazard planning at all stages: pre-disaster, during the disaster and post-disaster (WMO 2004).

Current thinking also promotes the engagement of the community to ensure local support and to formalize ownership of plans at the grass roots level. There is a widening recognition that community input is essential to any local disaster planning system since the community can provide valuable indigenous knowledge and are inevitably part of any warning system (Pilon 2002). This is not to the exclusion of regional- and national-level government planning, but in addition to them, and here it is seen as imperative that responsibilities between organizations are more clearly defined.

At the international level, cross-border collaboration is needed for trans-boundary catchments, which highlights the need for common methods of planning and sharing of expertise (Pilon 2002).

11.5 Conclusions

What this chapter set out to achieve was to emphasize to the reader that small island states such as those in the Caribbean are especially vulnerable to geohazards. In particular, it sought to highlight the fact that of all the geohazards in the region, floods are the most common and costly. The evidence presented illustrated that these facts have been known for several years, and yet still the consequences of flood events remain as acute as they ever were. Furthermore, the problem is set to grow since the impacts of global climate change and the associated sea level rise will foster conditions in which floods will be more frequent and more intense.

The challenges facing those responsible for managing flood risk in the Caribbean region are numerous. To start with, the region needs an observation network suitable for predicting the small and intense rainstorms that cause the most frequent flood events and which are supported by reliable and accessible technology. To interpret this data, the region must invest in the training of an array of specialist staff in a number of key academic disciplines, all of which contribute to the better understanding of floods and their causes.

Any investment in technology and training will have little impact if the outputs from these investments are not followed through, and in this respect there is a need for greater integration of organizations and policies. A particular problem associated with flooding is that catchments straddle administrative boundaries or even international borders, and actions taken in upstream areas will have consequences in downstream areas. This makes integration a necessary element of any effective flood risk management strategy. The lack of integration in these areas is not a problem unique to the Caribbean, but it reflects a challenge of global proportions. As such, many lessons can be learned from areas of the world where attempts have already been made to address this issue.

Conventional wisdom has dictated that solutions to flooding problems involve building ever stronger defenses or ever larger channels to cope with runoff. This structural approach is limited by the high capital cost incurred when implementing such schemes, and attention has been turning towards lower cost, non-structural ideologies. The non-structural paradigm embraces a holistic view of flood events, which seeks to incorporate not just integrated management, but also exploits the positive benefits floods may bring to the environment. IWRM is one dominant strategy that seeks to combine these two philosophies of structural and non-structural solutions and is gaining increasing support from around the world.

As a final note, the research associated with this chapter has revealed a common belief among researchers that to be effective, any flood risk management strategy must involve the community. The community is the ultimate beneficiary of any strategy; it possesses expert local knowledge that can aid flood prediction and forms an essential component of many non-structural measures. But community means many things to different people; it can be the people affected by floods, the organizations involved in the management of flood risk or the collection of states in a region. Whatever the definition used, a flood risk management strategy should never lose sight of the fact that its aim is to help and sustain the local community.

References

Baban, S.M.J., D. Thomas, K.J. Sant and F. Canisius. 2007. Managing development in the hillsides of Trinidad and Tobago. *Journal of Sustainable Development*. In press.

Baban, S.M.J., and R. Kantasingh. 2005. Mapping floods in the St Joseph Watershed, Trinidad, using GIS. *International Association of Hydrological Sciences (IAHS)*, publication no. 295:254–64.

Briguglio, L. 1995. Small island developing states and their economic vulnerabilities. *World Development* 23, no. 9:1615–32.

Caribbean Disaster Emergency Response Agency (CDERA). 2005. Caribbean Community regional programme framework 2005–2015. Paper presented at the Second World Conference on Disaster Reduction. Kobe, Japan.

Environment Agency (EA). 2001. *Lessons learned: Autumn 2000 floods*. Bristol, UK: EA.

———. 2003. *Strategy for flood risk management (2003/4–2007/8)*. Version 1.2. London: EA.

European Commission (EC). 2004. Flood risk management: Flood prevention, protection and mitigation. COM 2004, 472 final, Brussels.

European Environmental Agency (EEA). 2001. *Environmental issue report 21: Sustainable water use in Europe*. Part 3, Extreme hydrological events: Floods and droughts. Copenhagen: EEA.

Global Water Partnership Technical Advisory Committee (TAC). 2000. *TAC background paper 4:* Integrated Water Resource Management, Stockholm.

Hamill, L. 2001. *Understanding hydraulics*. 2nd edition. Basingstoke, UK: Palgrave.

Jones E.B., and B. Spence. 2003. The potential impacts of climate change and severe weather events on urban water resources in Jamaica: A case study for CDERA. Paper presented to the seminar Climate Change and Severe Weather Events in Asia and the Caribbean. Bridgetown, Barbados.

Laing, A.G. 2004. Cases of heavy precipitation and flash floods in the Caribbean during El Niño winters. *Journal of Hydrometeorology* 5:577–94.

Lancaster, J.W., M. Preene, C.T. Marshall. 2004. *Development and flood risk: Guidance for the construction industry.* London: CIRIA.

Mamat R., and S.B. Mansor. 1999. Remote sensing and GIS for flood prediction. *Proceedings of the Asian Association of Remote Sensing.* http://www.gisdevelopment.net/acrs/1999.

Melesse, A.M., and S.F. Shih. 2000. Geomorphic GIS database for runoff coefficient determination. *Proceedings of the Second International Conference on Geospatial Information in Agriculture and Forestry,* 505–12. Lake Buena Vista, Florida.

Ministry of Agriculture, Fisheries and Food and the Welsh Office (MAFF/WO). 1993. *Strategy for flood and coastal defence in England and Wales.* London: HMSO.

Nicholls, R.J., F.M.J. Hoozemans and M. Marchand. 1999. Increasing flood risk and wetland losses due to global sea-level rise: Regional and global analysis. *Global Environmental Change* 9:S69–S87.

National Rivers Authority (NRA). 1994. River Severn: Upper reaches catchment management plan. Consultation report. Shrewsbury, UK.

———. 1999. Planning and flood risk. Bristol, UK.

Pelling, M., and J.I. Uitto. 2002. Small island developing states: Natural disaster vulnerability and global change. *Environmental Hazards* 3:49–62.

Pilon, P.J. 2002. *Guidelines for reducing flood losses.* Geneva: UN-DESA.

Shih, S.F. 1996. Integration of RS and GIS for hydrologic studies. In *Geographical information systems in hydrology,* ed. V.P Singh and M. Fiorention, 15–42. Dordrecht: Kluwer Academic Publishers.

Wilson, E.M. 1983, *Engineering hydrology.* 3rd edition. London: Macmillan.

World Meteorological Organisation (WMO). 2004. *Water and disasters: Be informed and be prepared.* Geneva: WMO.

Xiuwan, C. 2002. Using RS and GIS to analyse land cover change and its impacts on regional sustainable development. *International Journal of Remote Sensing* 23, no. 1:107–24.

CHAPTER 12

Recognizing and Managing Unstable Slopes in Trinidad and Tobago

SERWAN M.J. BABAN and JOHN B. RITTER

Abstract

Landslides tend to have a serious impact on the environment, population and the economies of small mountainous tropical islands. This is mainly due to the existing favourable physical conditions for landslides and their triggers, coupled with the increasing demands of development, tourism and population growth.

This chapter aims to develop a broad field-based method to recognize unstable slopes and to provide some simplified options for mitigating this instability. This objective is being realized by presenting three ways of developing GIS-based slope-stability approaches for tropical mountainous environments through examining Trinidad. The approaches are as follows: overlay analysis using variables related to soil/geologic combinations or derivatives, a probability model based on potential influencing variables and a physically based deterministic model using map algebra. The sources of raw data, including soil, geology and digital terrain models (DTM), are identified, and techniques to capture, validate and convert the required data into usable digital format for analysis are described. The outcomes are represented and evaluated based on field evidence.

12.1 Introduction

Landslides are a part of the natural processes of terra formation, occurring both naturally and as a result of human interactions with the environment. Landslides can result in loss of property, loss of opportunity and loss of life. Although not all landslides result in catastrophe, the damage from many small landslide events can add up and exceed the impact of a single major failure.

The combination of steep slopes and high-intensity rainfall leads to major landslide problems in areas such as the Northern Range of Trinidad and the Main Ridge area of Tobago. These areas have limited access routes, and landslide events have severe impacts at the social as well as economic levels. On the island of Trinidad, there are three mountain ranges, the most significant being the Northern Range. The Central and Southern Ranges, which may be described as continually undulating hills, are separated by the Naparima Plains. The flood plain of the Caroni River, located at the southern side of the Northern Range, is quite flat, and landslides have little impact on it. The Central Range exhibits significant discontinuities within its geological structure, with heavy and sensitive clays occurring over wide areas. The Southern Range is the smallest of the three ranges and also has significant discontinuities within its geological structure, with clays and marls occurring over large areas (Figure 6.2).

Landslides occur most frequently in the phillite mica-schist, sandstone and clay lithologies (Bertrand et al. 1986). In mountainous areas, a relatively shallow soil layer, and in several cases exposed weathered bedrock material, results in the propensity for rock and soil falls and topples, with translational slides and debris flow being the predominant landslide types (Baban and Sant 2004). As one moves towards areas with clays and marls, there is a change in landslide types to lateral spreading and flows, with complex slides and rotational slides being the most prevalent types. The reactivation of slides has become evident in mountainous areas, as slides repeatedly occur annually in the same location (Baban and Sant 2005). For example, on the North Coast Road in Trinidad, landslides occurring during 2003, 2004 and 2005, were mapped and found to have a high level of re-occurrence, leading to the conclusion that these slides are active. Similarly, the toes of rotational slides, in the central southwest of Trinidad, move each year during the wet season and remain relatively inactive during the dry season. In December 2004, the North Coast Road was closed from the public for a period of one week due to the occurrence of scores of landslides that rendered the road impassable.

Trinidad and Tobago lies within the tropics and experiences a tropical marine climate. Tropical cyclones or hurricanes provide an abundance of opportunities for high-intensity precipitation events. Furthermore, altitude influenced convention rainfall events can also provide both long duration and high-intensity rainfall events, which are regarded as primary landslide triggers in the Caribbean. Anthropogenic activities and seismic events are also significant contributors towards the occurrence of landslides (Baban and Sant 2005).

There have been few geotechnical studies into slope stability. In those cases where geotechnical studies have been done for major infrastructure works, their distribution is usually concentrated to the urban areas, and the sampled soil parameters/properties are inconsistent. The absence of a complete soil parameter database that provides information on all soils has restricted the development of numerical modelling for slope-stability studies, except in cases where there has been a significant impact on local infrastructure. In the current circumstances, there are no official susceptibility studies on the regional scale for landslide susceptibility.

The absence of a landslide recording system has impeded the investigation of landslide susceptibility in many Caribbean island states (Ahmad 2003; Baban and Sant 2004). This in turn has had a negative influence in the formation of proper national planning policies and property insurance systems. In the worst-case scenarios, developments are permitted on slopes prone to failure, without adequate slope stabilization works, and homeowners are often made to pay disproportionate insurance premiums based on an average figure, as opposed to the actual degree of landslide vulnerability to which their properties may be subjected. Consequently, there is a need to identify general conditions related to unstable slopes based on field observations.

This chapter aims to develop a broad field-based method to recognize unstable slopes and to provide some simplified options for mitigating the risk.

12.2 Conditions Associated with Unstable Slopes

Landslides occur as a result of a number of determining and triggering factors. Therefore, landslide susceptibility analysis will invariably require the identification and quantification of these factors (Varnes 1978).

General conditions related to unstable slopes lead to landslides and are both intrinsic to the slope (for example, geological causes) or extrinsic (for example, morphological, physical or human causes). The extrinsic causes generally trigger landslides on slopes that are predisposed to slope instability through their intrinsic properties (Table 12.1).

Table 12.1 Checklist of Landslide Causes

1.	**Geological Causes**	
a.	Weak materials	√
b.	Sensitive materials	√
c.	Weathered materials	√
d.	Sheared materials	√
e.	Jointed or fissured materials	√
f.	Adversely oriented mass discontinuity (bedding, schistosity, etc.)	√
g.	Adversely oriented structural discontinuity (fault, unconformity, contact, etc.)	√
h.	Contrast in permeability	√
i.	Contrast in stiffness (stiff, dense material over plastic materials)	√
2.	**Morphological Causes**	
a.	Tectonic or volcanic uplift	√
b.	Fluvial erosion of slope toe	√
c.	Wave erosion of slope toe	√
d.	Erosion of lateral margins	
e.	Subterranean erosion (solution, piping)	√
f.	Deposition loading slope or its crest	√
g.	Vegetation removal (by forest fire, drought)	√
3.	**Physical Causes**	
a.	Intense rainfall	√
b.	Prolonged exceptional precipitation	√

Table 12.1 continues

Table 12.1 Checklist of Landslide Causes *(cont'd)*

c.	Rapid drawdown (of floods and tides)	√
d.	Earthquake	√
e.	Shrink-and-swell weathering	√
4.	**Anthropogenic Causes**	
a.	Excavation of slope or its toe	√
b.	Loading of slope or its crest	√
c.	Drawdown (of reservoirs)	√
d.	Deforestation	√
e.	Irrigation	√
f.	Mining	
g.	Artificial vibration	
h.	Water leakage from utilities	√

Note: The √ indicates causes that may be especially pertinent to or pervasive in Trinidad and Tobago.
Source: Cruden and Varnes 1996.

12.2.1 Geological Conditions

Geological causes of landslides are basically related to rock type, altitude of bedding (strike and dip) and orientation of other discontinuities such as faults, fractures and joints, and foliation. Figure 12.1 illustrates folded, interbedded tan and brown sandstones and shales (DeGraff et al. 1989). The shales are thin and comprise the bedding plane between several of the sandstone layers. The shale beds are expressed by the increased vegetation along them, suggesting they have sufficient water-holding capacity to support plant growth. Incompetent rock types, such as shale, phyllite or schist, have lower shear strength than more competent rock types such as massive sandstone or limestone. Unconsolidated or weakly consolidated clay, silt, sand and gravel are less competent than their consolidated counterparts (Wharton 1994). In interbedded lithologies, bedding planes separate rock types of different competence (Kugler 1961). In addition, the incompetent beds, though of negligible thickness, may comprise the failure plane for a landslide. Orientation of bedding is particularly critical when the strike of bedding planes is par-

Figure 12.1 Folded, inter-bedded sandstones and shales. The arrows show dip directions of bedding.

Figure 12.2 Strike and dip of bedding relative to hill slope and aspect affect slope stability. In this particular case, the overlying unconsolidated material was removed because slope failures repeatedly occurred along the exposed bedding plane.

allel to hillside, and beds dip in the direction of slope and at a similar degree of slope (Figure 12.1).

Rock strength is least along a discontinuity, and as a result, the discontinuity becomes the failure plane. In Figure 12.2, the bedding plane surface shown represented a critical discontinuity between the overlying colluvial soil and weathered material and the underlying bedrock; the discontinuity was the failure plane along with sliding, which probably repeatedly occurred until the weathered mantle was removed.

Strike of bedding in this scene parallels the trend of the hill slope and the road at the base of the hill slope. The dip of the beds points down slope and parallels the surface slope. Competent and incompetent rock types can be made less competent over geologic time by faults, fractures and joints (Figure 12.3). In addition, foliation resulting from metamorphism presents an additional discontinuity that is susceptible to failure for some of the same reasons. The strike and dip of beds, faulting, and foliation may be represented on geologic maps. Map symbols represent

Figure 12.3 Foliation and fracturing of phyllitic schist in this outcrop are high angle and essentially parallel to each other. Only two fracture orientations are shown, but there are several more, including one nearly parallel with the face of the outcrop. Any one of these discontinuities may be a potential failure plane, especially when aligned or congruent with morphological or human causes of slope instability.

regional orientations and relationships, but localized irregularities should be evaluated in the field.

In other cases, the failure may be the result of differences in permeability, or material properties between or within units. The debris flow in Figure 12.4a resulted from failure, not at the interface between the soil and underlying geology, but rather within the underlying geology – a grey plastic clay unit (Figure 12.4b). Strata within the clay unit may

Figure 12.4a A debris flow is characterized by flow within the slide mass as indicated here by the expanding front.

Figure12.4b The failure plane occurred within a viscous clay unit underlying the soil as exposed by headwall scarp of this flow.

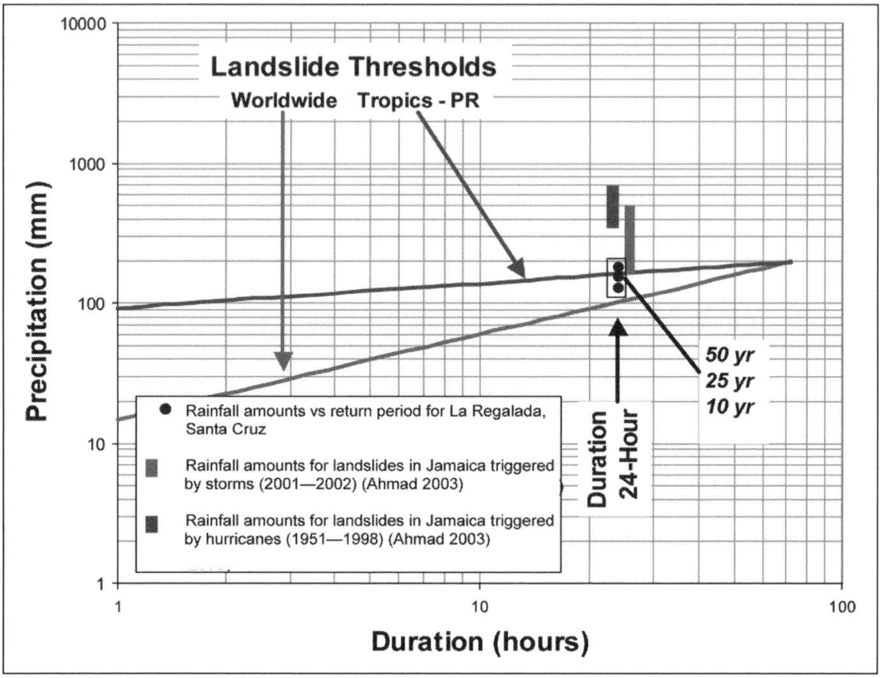

Figure 12.5 Rainfall amount and duration have been used to establish thresholds for conditions that may result in widespread landslides. The worldwide and tropical thresholds are based on the work of Caine (1980) and Larsen and Simon (1993) respectively. This is a summary of 24-hour precipitation amounts for different types of storm events, including frontal-type events of different return periods for the La Regalda station (#9.52) in Santa Cruz, Trinidad, and events observed to cause widespread landslides in Jamaica are also shown.

contain layers of more expansive or more plastic clay, which would deform more easily or less permeable clay, which would cause an increase in pore pressure.

12.2.2 Morphological Conditions

Morphological causes of slope instability are external to the hillslope system and work over variable spans of time, from geologic intervals of time or much shorter time intervals (for example, a flood event). Morphological causes of slope instability increase the driving force, or shear stress acting on a slope, either by increasing slope through uplift or by removal of lateral support along the toe of the slope (DeGraff et al. 1989; Wharton 1994) (Table 12.1).

12.2.3 Physical Conditions

The physical condition that causes most landslides in Trinidad and Tobago is excessive precipitation, particularly during wetter than normal wet seasons. Figure 12.5 summarizes the rainfall-duration conditions under which landslides have been known to occur worldwide and in tropical climates (Aliasgar and Baban 2006).

12.4 Human Conditions

Humans impact slope stability by increasing shear stress on the slope or decreasing shear strength of the slope. Excavating along a slope, or at the toe of the slope, increases the shear stress on the slope (Wharton 1991a, 1991b). This is particularly critical where excavation causes significant discontinuities to "daylight" along the excavated face (Figure 12.6). Loading the slope or slope crest through filling or building structures also increases the shear stress. Other human influences reduce the shear strength by changing the pore water pressure (for example, leak-

Figure 12.6 Exposing potential failure surfaces through excavation, particularly where they dip toward the excavation, decreases slope stability.

ing water mains) (Baban and Sant 2005). Deforestation may change the shear strength by decreasing the cohesiveness caused by roots, but it also may increase the amount of water within the soil mantle, which also acts to reduce shear strength.

12.5 Recognition of Unstable Slopes

While the conditions that cause or contribute to landslides may be evaluated in the field or from maps, recognizing slopes that are actually unstable, or potentially unstable, in the field is more difficult. Landscape features that characterize slope instability in the field or, if recent, from aerial photographs, are summarized in Table 12.2. These features are based on morphologic, vegetation and hydrologic conditions that result from previous landslides, emphasizing the point that previously unstable slopes may be unstable in the future. Materials that were involved in a previous slide may continue to slide lower on the slope or, as addi-

Table 12.2 Morphologic, Vegetation and Drainage Features Characteristic of Slope Instability

Terrain Feature	Relation to Slope Instability
Morphology	
Concave/convex slopes	Landslide niche and associated deposit
Step-like slopes and tension cracks	Retrogressive sliding
Semicircular scarps and steps	Head part of slide with outcrop of failure plane
Hummocky and irregular slope	Microrelief associated with shallow movements or small retrogressive slide blocks; also typical of old landslide deposits
Infilled valleys with slight convex bottom, where v-shaped valleys are normal	Mass movement deposit of flow-type form

Table 12.2 continues

Table 12.2 Morphologic, Vegetation and Drainage Features Characteristic of Slope Instability *(cont'd)*

Terrain Feature	Relation to Slope Instability
Vegetation	
Vegetational clearances on steep scarps, coinciding with steps	Absence of vegetation on headscarp or on steps in slide body
Irregular linear clearances along slope	Slip surface of translational slides and track of flows
Disrupted, tilted, disordered or partly dead vegetation	Slide blocks and differential movements in body
Differential vegetation associated with changing drainage conditions conditions on body	Ponded drainage on back-tilted blocks, seepage at frontal lobe and differential
Hydrology	
Ponded drainage or sag ponds	Landslide niche, back-tilting landslide blocks and hummocky internal relief on landslide body
Seepage and spring levels	Springs along frontal lobe and at places where failure plane outcrops
Mottled or gleyed soils	Seasonal high groundwater table may be above potential failure planes

Source: Sidle et al. 1985; Soeters and van Westen 1996.

tional debris collects on the slope, it may become unstable and slide on the same failure plane.

Several features associated with soil creep may be indicative of a slope that is moving. Creep is a term with multiple meanings. It refers to a type of soil mass wasting process, resulting from cyclic expansion and contraction of the soil mantle, by wetting and drying of the soil and shrinking and swelling of clays within the soil. It also refers to a rate of slope movement (for example, very slow to extremely slow, < 1.6 m/year according to Cruden and Varnes [1996]). Recognizing landscape features that are characteristic of creep is critical, because it may represent continued movement on an advancing or enlarging landslide, or on a slope condition that over time will become increasingly less stable, because of over steepening. Figures 12.7–12.10 illustrate several indicators of creep.

Figure 12.7 Tension cracks on this highway fill indicate downslope movement of the fill material. Foliations and fractures in the bedrock along the right side of the photo are oriented in the same general direction.

Figure 12.8 The convex out-slope of both the actual surface and the top of the vegetation is an indicator of progressive creep downslope. This movement is occurring along a potential failure plane and may result in slope failure because of oversteepening.

Figure 12.9 The tilting telephone poles are evidence of progressive creep towards the road. Similar orientations of trees are an additional indicator as are taut wires when utility poles are oriented in a downslope direction.

Figure 12.10 Continuous creep as indicated by the failure and cracks in the retaining wall.

Table 12.3 Non-structural Approaches to Potential Slope-stability Problems

Approach	Objective	Method or Procedure
Avoid the problem	Restrict development in landslide-prone areas by relocation of existing development	Relocation of existing development through public acquisition of land and conversion of existing structures to uses less vulnerable to slope failure
	Restrict development in landslide-prone areas by discouraging new development	Information programmes to educate developers and potential purchasers, disclosure of hazards to potential property purchasers, exclusion of public utilities, tax credits, financing policies and insurance costs, and legal liability
	Regulate future development	Regulating future development through land-use zoning regulations, subdivision regulations and sewage-disposal regulations
Don't exacerbate the problem	Regulate excavation, construction, and grading	Establish building codes, possibly in overlay districts, requiring permits before slopes are scraped, excavated, filled, or cut; regulate, minimize or prohibit excavation and fill; control disruption of drainage and vegetation
Reduce driving forces or increase resisting forces	Maintain or conserve natural cover and regulate re-vegetation of developed sites	Establish policies and educate developers and purchasers of the importance of water to slope stability; use natural vegetation to reduce soil moisture, lower regional or perched water tables and provide cohesion
	Reduce load on slope	The load on slopes can be reduced by subdivision regulations prohibiting high-load types of structures like swimming pools; zoning regulations can specify critical slopes where fill cannot be placed

Table 12.4 continues

Table 12.4 Structural Approaches to Potential Slope-stability Problems *(cont'd)*

Approach	Objective	Method or Procedure
Divert or buffer the problem	Contain or direct land-slide debris from critical facilities	Benches or terraces constructed along unstable slope should be graded to divert water off the slope; deflection walls with up-slope drainage
Reduce driving forces	Drain surface water away from critical slopes	Reduce weight of the potential slide mass by decreasing water weight; re-grade to divert surface water from critical slopes; establish brow ditches on upper slopes to decrease water on lower slopes
	Drain subsurface water away from critical slopes	Reduce weight of the potential slide mass by decreasing water weight; construct trench drains or interceptor drains to remove subsurface water from above critical slopes; construct drains behind barrier structures
	Reduce grade of slope	Combination of barrier structure and backfill to lower the grade of slopes
Increase resisting forces	Apply external force	Use buttress and counterweight fills; toe beams
	Apply external force and modify grade	Use structural barriers with externally (cantilever, gravity, braced, tied-back) or internally (reinforce soil, soil nailing) stabilized systems
	Create adhesion between slide mass and potential failure plane	Install tieback anchors, ground anchors, soil nails or rock bolts

Table 12.4 continues

Table 12.4 Structural Approaches to Potential Slope-stability Problems *(cont'd)*

Approach	Objective	Method or Procedure
Increase internal strength	Drain subsurface to reduce pore pressure	Construct trench drains or interceptor drains to remove subsurface water from above critical slopes; construct drains behind barrier structures
	Use reinforced backfill or install in-situ reinforcement to resist shear	Backfill materials are reinforced using steel strips, welded wire sheets, bar mats and meshes, geotextiles, geo-grids, and fibres

12.6 Mitigating Slope Instability

The US Geological Survey (1982) has identified four factors that underpin successful landslide hazard-mitigation programmes in the United States (Schuster and Kockelman 1996):

1. An adequate base of technical information on the hazards and risks
2. A technical community able to apply and enlarge upon this data base
3. An able and concerned local government
4. A citizenry that realizes the value of and supports a programme that promotes health, safety and the general welfare of the community

Structural and nonstructural forms of mitigation will be introduced here, but the particulars of any mitigation method must be tailored to address local conditions and meet community needs (Tables 12.3 and 12.4). Particularly with regard to mitigation efforts, and especially structural methods, the solutions are engineered and require geo-technical analysis of the site as the proposed method of mitigation. Because water is usually the trigger of landslides, managing drainage on potentially unstable sites, is essential for reducing the risk of landslides. Holtz and Schuster (1996) note that appropriate drainage of surface and sub-surface water is the most widely used and generally the most successful slope stabilization method because of its high stabilization efficiency in relation to design and construction costs.

References

Ahmad, R. 2003. Developing early warning systems in Jamaica: Rainfall thresholds for hydro-geological hazards. Paper presented at the Office of Disaster Preparedness and Emergency Management National Disaster Management Conference. Ochos Rios, Jamaica. http://www.mona.uwi.edu/uds/rainhazards_files/frame.htm.

Aliasgar, K., and S.M.J. Baban. 2006. Developing a geoinformatics based early warning system for landslides in Tobago. *Proceedings of Urban and Regional Information Systems Association (URSIA) Conference*. The Bahamas.

Baban, S.M.J., and K.J. Sant. 2004. Mapping landslide susceptibility on a small mountainous tropical island using GIS. *Asian Journal of Geoinformatics* 5, no. 1:33–42.

———. 2005. Mapping landslide susceptibility for the Caribbean island of Tobago using GIS, multi-criteria evaluation techniques with a varied weighted approach. *Caribbean Journal of Earth Sciences* 38:11–20.

Bertrand, D., H. Romano and C.T. Rogers. 1986. Landslide and flood distribution in the west coastal area of Trinidad: The role of geology. *Transactions of the First Geological Conference of the Geological Society of Trinidad Tobago*, 129–39.

Caine, N. 1980. The rainfall intensity-duration control of shallow landslides and debris flows. *Geografiska Annaler*, ser. A, Physical Geography, 62:23–27.

Cruden, D.M., and D.J. Varnes 1996. Landslide types and processes. In *Landslides: Investigation and mitigation*, ed. A.K. Turner and R.L. Schuster, 36–75. Transportation Research Board Special Report, no. 247. Washington, DC: National Academies Press.

DeGraff J.V, R. Bryce, R.W. Jibson, S. Mora and C.T. Rogers. 1989. Landslides: Their extent and significance in the Caribbean. In *Landslides: Extent and economic significance*, ed. E.E. Brabb and B.L. Harrod, 51–80. Rotterdam: Balkema.

Holtz, R.D., and R.L. Schuster. 1996. Stabilization of soil slopes. In *Landslides: Investigation and mitigation*, ed. A.K. Turner and R.L. Schuster, 439–73. Transportation Research Board Special Report, no. 247. Washington, DC: National Academies Press.

Kugler, H.G. 1961, *Geological map and seven sections/1:100,000 of Trinidad*. Zurich: Orell Fussli.

Larsen, M.C., and A. Simon. 1993. A rainfall intensity-duration threshold for landslides in a humid tropical environment in Puerto Rico. *Geografiska Annaler,* ser. A: Physical Geography, 75:13–23.

Schuster, R.L., and W.J. Kockelman. 1996. Principle of landslide hazard reduction. In *Landslides: Investigation and mitigation*, ed. A.K. Turner and R.L. Schuster, 91–105. Transportation Research Board Special Report, no. 247. Washington, DC: National Academies Press.

Sidle, R.C., A.J. Pearce and C.L. O'Loughlin. 1985. *Hillslope stability and land use*. Water Resources Monograph, no. 11. Washington, DC: American Geophysical Union.

Soeters, R., and C.J. van Westen. 1996, Slope instability recognition, analysis, and zonation. In *Landslides: Investigation and mitigation*, ed. A.K. Turner and R.L. Schuster, 129–77. Transportation Research Board Special Report, no. 247. Washington, DC: National Academies Press.

US Geological Survey (USGA). 1982. *Goals and tasks of the landslide part of Ground-failure Hazards Reduction Program*. US Geological Survey Circular no. 880. Washington, DC: USGA.

Varnes, D.J. 1978. Slope movement type and processes. In *Landslides: analysis and control,* ed. R.L. Schuster and R.J. Krizek, 20–47. Transportation Research Board Special Report, no. 176. Washington, DC: National Research Council.

Wharton, S.R. 1991a. Regional evaluation of cut-slope instability using statistical methods, Northern Range, Trinidad. *Proceedings of the Second Annual Meeting of the Geological Society of Trinidad and Tobago*, 107–17.

———. 1991b. Cut-slope instability and associated sub-watershed development, Carenage/Chaguaramas, Trinidad. *Proceedings of the Second Annual Meeting of the Geological Society of Trinidad and Tobago*, 118–30.

———. 1994. Landslide hazard analysis in hilly tropical terrain in Trinidad and Tobago. *Proceedings of the Caribbean Conference on Natural Hazards: Volcanoes, Earthquakes, Windstorms, Floods*, ed. W.B. Ambeh. St Augustine, Trinidad.

CHAPTER 13

Developing Early Warning Systems for Managing Geohazards in the Caribbean

SERWAN M.J. BABAN and KELLY ALIASGAR

Abstract

Events such as landslides and floods can be difficult to manage, and statistics show there is an urgent need to mitigate associated risks. These hazards not only damage homes, livelihoods and critical resources as well as take lives, but they have devastated the economies of the affected islands. The occurrence and the impacts of these natural hazards are magnified in the Caribbean islands due to the presence of suitable conditions: size, location, geological structure and, more importantly, the lack of hazard-mitigating strategies. Increased deforestation and development activities exacerbate the situation, increasing the possibility of these types of geohazards. Clearly, there is a need to manage and predict these disasters in order to minimize risk and losses. Geoinformatics can assist with managing geohazards through developing inventory maps and identifying past events, as well as mapping and analysing favourable conditions and triggers.

This chapter attempts to manage this problem through developing a geoinformatics-based early warning system for landslides and floods in the Caribbean region by using Tobago as a case study. The system is based on monitoring rainfall patterns and developing threshold values for rainfall intensities and durations, which can trigger landslide and flood events. These thresholds can then be analysed with susceptibility maps that can be used to predict the onset of geohazards. Areas at risk can then be identified, and evacuation routes can be determined before the disaster even occurs. Early warning systems have been proven successful in many countries, and implementation and use in the Caribbean can be advantageous.

13.1 Introduction

The use of early warning systems has been considered a cornerstone for disaster reduction. The United Nations International Decade for Disaster Preparedness and Management has increased awareness of the need for early warning systems and its critical role in minimizing, if not preventing, the hazardous effects of disasters. The lack of early warning systems is far too common, and many communities and countries that are highly vulnerable to natural hazards are still without such systems. The effects of not having such a predictive system were clearly demonstrated by the destruction caused by the Caribbean hurricanes of 2004 (ISDR 2006).

Early warning systems can be defined as “the provision of timely and effective information, through identifying institutions, that allow individuals exposed to a hazard to take action to avoid or reduce their risk and prepare for effective response” (Glantz 2003). The existence of an effective early warning system can empower individuals and communities threatened by hazards to act in sufficient time and in an inappropriate manner so as to reduce the possibility of personal injury, loss of life and damage to property or the environment. These systems tend to have components dealing with prior knowledge of risk, a technical monitoring warning service, communicating the risk and preparing communities to act (PPEW 2006) (Table 13.1).

In order for an early warning system to be successful it will need to possess all of the following characteristics of early warning systems (Helms 1989; Alexander 1995; Glantz 2003; ISDR 2004; Aliasgar and Baban 2006):

Table 13.1 Basics of Early Warning Systems

Prior knowledge of the risks faced by communities	**Technical monitoring and warning service for these risks**	**Dissemination of understandable warnings to those at risk**	**Knowledge and preparedness to act**
Risks arise from both the hazards and the vulnerabilities that are present – what are the patterns and trends in these factors?	Is there a sound scientific basis for predicting the risks faced? Are the right things being monitored? Can accurate warnings be generated in a timely fashion?	Do the warnings reach those at risk? Do people understand them? Do they contain useful information that enables proper responses?	Do communities understand their risks? Do they respect the warning service? Do they know how to react?

1. Continuity in operation: The system needs to be able to identify and monitor potential dangers as early as possible. Thus, it must operate 24 hours a day, year round.
2. Clear scientific basis: The system must be developed based on clearly identified triggers and thresholds.
3. Stability: The system should be able to function under severe conditions and to physically withstand the impact of possible disasters.
4. Adequate warning time: A lack of warning time will defeat the purpose of the system. Although there are substantial differences in warning times, for example, between droughts, which may provide months of warning, and tornados, which may provide minutes of warning, there should be efficient warning time for actions and responses to be activated.
5. Transparency: All involved must clearly understand the process and ensure the system is not biased to sponsors, political parties or privileged groups.
6. Integration: The system should be an integral part of the larger systems, whether community oriented, government oriented or culturally oriented.

7. Realistic and capable of being implemented: The system should be based on available physical and human resources, both locally and nationally.
8. Flexible: The system should contain some aspect of capacity planning to cater for expanding systems and/or new indicators.
9. Apolitical: The system should be driven by science and engineering without any political interference from any groups in society.

Furthermore, some of the main principles and responsibilities for effective early warning systems include the following:

1. Awareness by officials and the vulnerable population of the hazards to the related effects they are exposed to and the necessary actions that will minimize the threats and loss of damage;
2. Community leaders should understand the advisories received and be able to advise, instruct and lead the population in a manner, which will not increase safety and reduce damages and losses; and
3. In addition to the above principles, the structure of the advisories and warnings should be one that considers the cultural and social aspects of the society and is comprehensible and accessible to all affected and potentially affected. It is important that the population does not under- or overestimate the intended severity of the warning.

Finally, a critical link to the early warning chain is the response to the warning itself and, more crucially, the public's reaction in these situations. This aspect is probably most crucial, as the appreciation of psychological, community and individual processes in stressful times is more important than technology. Some critical components to ensure the best public response to the warnings include the following specific elements (ISDR 2004):

1. Sufficient lead time
2. Accuracy
3. Understanding and belief in the warning
4. Understanding the reality of a threat
5. Confirmation of the warning from other sources
6. Knowing how to react
7. Being prepared

13.2 Rainfall, Hydrological Triggers and Thresholds

Rainfall is defined as "the amount of precipitation of any type, primarily liquid. It is usually the amount that is measured by a rain gauge" (NOAA 2005). The Caribbean region is part of the humid tropics, which is defined as those zones with tropical rainy climates on either sides of the equator and includes the lowlands of the river Amazon, the central Congo basin, the African Guinean coast, the southern Asiatic peninsula, Malaysia, Indonesia, the Philippines, the Brazilian Atlantic coast, Madagascar, Bangladesh, Northern India, Central America and the Caribbean Islands (Keller 1983). Although these land masses are of varying sizes and of different geological and topological structures, they share similarities in their rainfall and weather. There are two common characteristics of all the Caribbean islands that affect rainfall (Macpherson 1984): (1) rain falls in heavy showers (as a result of the temperature of land), and (2) there are two main seasons: wet and dry.

In most of the islands in the Antilles, tropical storms and depressions result in some of the most intense rain, as opposed to non-cyclonic rains, and although both cyclonic and non-cyclonic rains cause floods in the Caribbean, the latter results in more devastating floods (Arenas 1983a). Adding to this, the basins of the Greater and Lesser Antilles are relatively small, and the rivers are short with steep slopes (Arenas 1983b). Increased rainfall is one of the main causes of flooding not only in the Caribbean but worldwide. By monitoring rainfall as a trigger for floods, it is possible to predict the occurrences of these hazards and issue early warnings (Aliasgar and Baban 2006).

By its nature, and similar to floods, rainfall has both spatial and temporal characteristics, as it occurs at a particular place at a certain time and for a duration of time. Because of this, it is possible to monitor the effects of rainfall in the Caribbean. However, in order to predict floods, the threshold values of sensitive areas need to be established. Examining the hydrological cycle will indicate that there are many variables affecting the rate at which rainfall runs off into rivers, streams and other waterways.

When precipitation enters the soil, it is absorbed until the maximum storage capacity of the soil has been reached. This quantity is dependent on the soil moisture content before precipitation and the rate of evaporation and transpiration. When maximum infiltrative capacity is reached, surface runoff occurs (Gumbricht 1996). The following list outlines the factors that affect the rate of runoff (USGS 2005):

1. Meteorological factors affecting runoff:
   - Type of precipitation (rain, snow, sleet, etc.)
   - Rainfall intensity
   - Rainfall amount
   - Rainfall duration
   - Distribution of rainfall over watersheds
   - Direction of storm movement
   - Antecedent precipitation and resulting soil moisture
   - Other meteorological and climatic conditions that affect evapotranspiration, such as temperature, wind, relative humidity and season
2. Physical characteristics affecting runoff:
   - Land use
   - Vegetation
   - Soil type
   - Drainage area
   - Basin shape
   - Elevation
   - Slope
   - Topography
   - Direction of orientation
   - Drainage network patterns
   - Ponds, lakes, reservoirs, sinks and so on in the basin, which prevent or alter runoff from continuing downstream

Runoff can be monitored by the use of stream flow gauges, where the volume of water in the waterways can be monitored. Once the maximum capacity of these waterways and the rate of runoff are established, the occurrence of floods can be predicted. Hydrographs can be used to compare the changes in the stream flow when rainfall increases. Hydrographs are graphs showing the water level (stage), discharge or other property of a river volume with respect to time (USGS 2005). The hydrograph in Figure 13.1 illustrates the increases in stream flow as the amount of rainfall increases. These graphs can be used in determining the onset of floods once threshold values are known.

Using hydrometerological data for expected rainfall, the rate of runoff can be determined, hence, it is possible to calculate when threshold values for river basins and other waterways are approaching. Flood-hazard maps can then be used to determine the affected areas (Aliasgar and Baban 2006).

In order to create early warning systems with rainfall, a trigger threshold value needs to be created. As such, there needs to be the continuous collection of hydrological and meteorological data for all catchments. In addition, historical data indicating the past threshold values should have been collected to give an indication of the combinations which triggered floods. Other sources of data needed would be as follows:

1. Catchment areas
2. Soil type
3. Land use/land cover

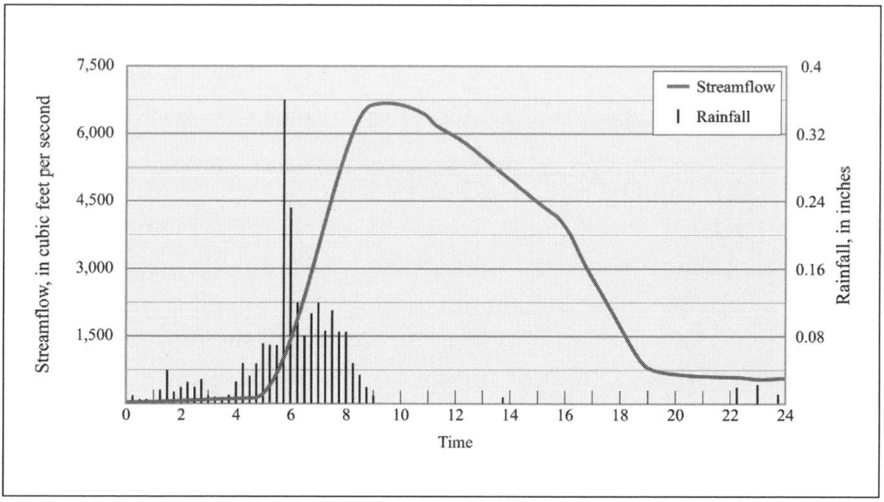

Figure 13.1 Hydrograph illustrating the increases in stream flow as the amount of rainfall increases.

4. Impervious surface (water penetrating)
5. Elevation (slope, slope length and aspect)
6. Rainfall data (duration, wind directions, intensities)
7. Stream flow data (volume)

Remote sensing techniques have been used by many countries to collect temporal and spatial data on rainfall to be used in determining threshold values for early warning systems. Three of the most common methods used involve the use of radar, satellite and rain gauges.

13.2.1 Developing Threshold Values

Once rainfall data has been collected, the analysis and trends of rainfall induced flooding and landslides can be assessed. In order to predict floods and landslides, the threshold values of sensitive areas need to be established (Aliasgar and Baban 2006).

There are many variables affecting the rate at which rainfall runs off into rivers, streams and other waterways. Threshold values depicting rainfall intensity and durations have and are being used by many countries in forecasting hydrometerological geohazards. Thresholds can be defined as minimum (lower boundary, below which no landslide will occur) or maximum (above which there is 100% probability occurrence of landslides) (LessLoss 2005). A rainfall value that lies between these two thresholds has varying probabilities of hazard occurrence. As the rainfall values approach the maximum thresholds, the probability of a hazard occurring increases (Glade et al. 2000).

Essentially, rainfall amounts before the landslides and floods are plotted to determine the combination of rainfall intensity and duration that triggers the flood or landslide. When these thresholds have been reached or are fast approaching (this can be determined from meteorological predictions), communities can be warned of the possible onset of floods or landslides. Rainfall thresholds can be developed, and have been developed, through empirical, statistical and physical methods. Using historical rainfall data for landslides that did and did not occur develops empirical thresholds. The result is the minimum threshold values (LessLoss 2005). Figure 13.2 displays some worldwide intensity/duration models.

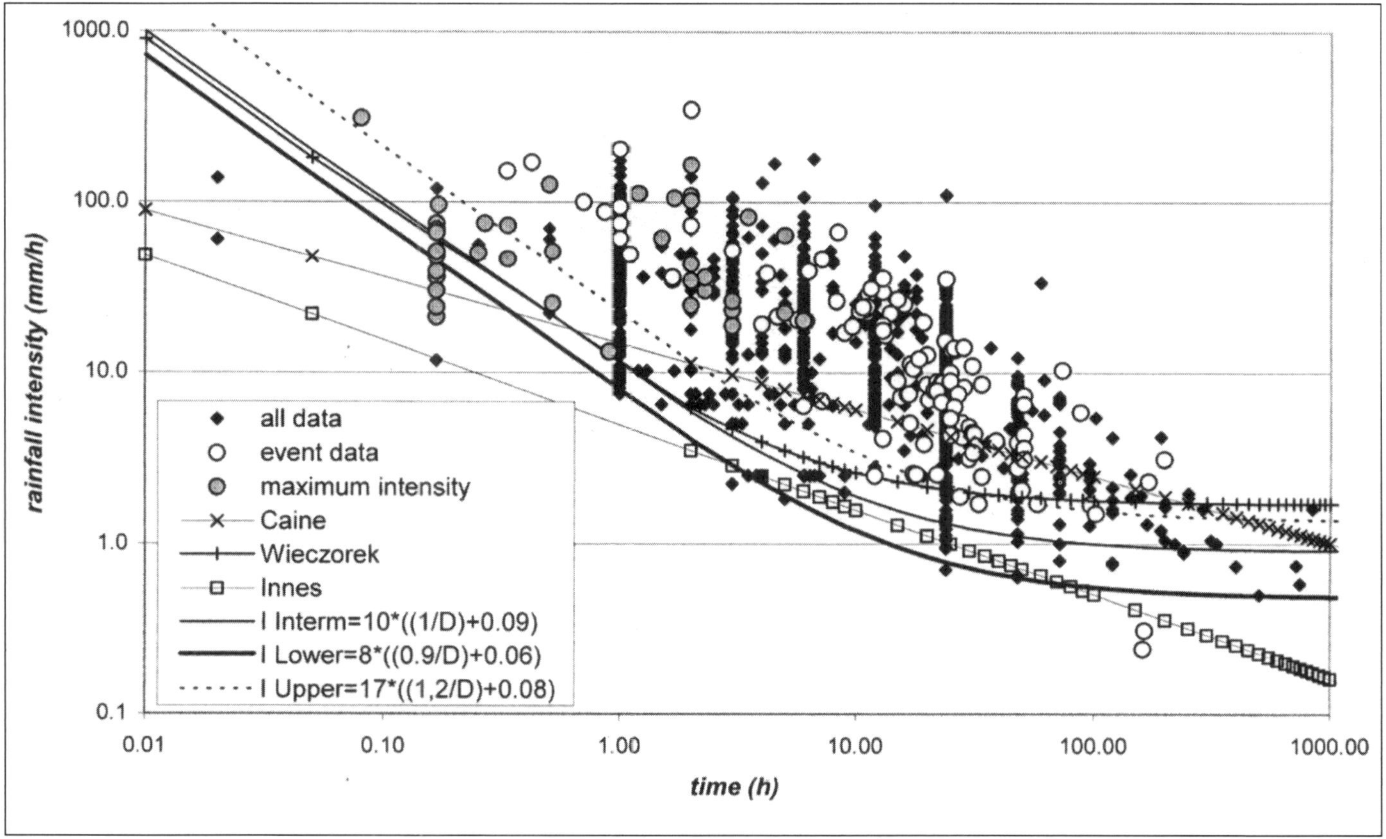

Figure 13.2 Global empirical thresholds (LessLoss 2005).

13.2.2 Caribbean Case Studies

Jamaica

Jamaica has an annual rainfall of approximately 125 cm to 700 cm, and two thirds of the area is on mountainous zones. Hourly data for 23 storms between 1951 and 2002 have been used to develop two rainfall intensity/duration thresholds for landslides.

The threshold developed indicates that for rainfalls of short duration (about 1 h), higher than 36 mm/h is required to trigger landslides (Figure 13.3). Low average intensities of about 3 mm/h appear to be sufficient to cause land sliding, as storm duration approaches approximately 100 h (Ahmad 2003).

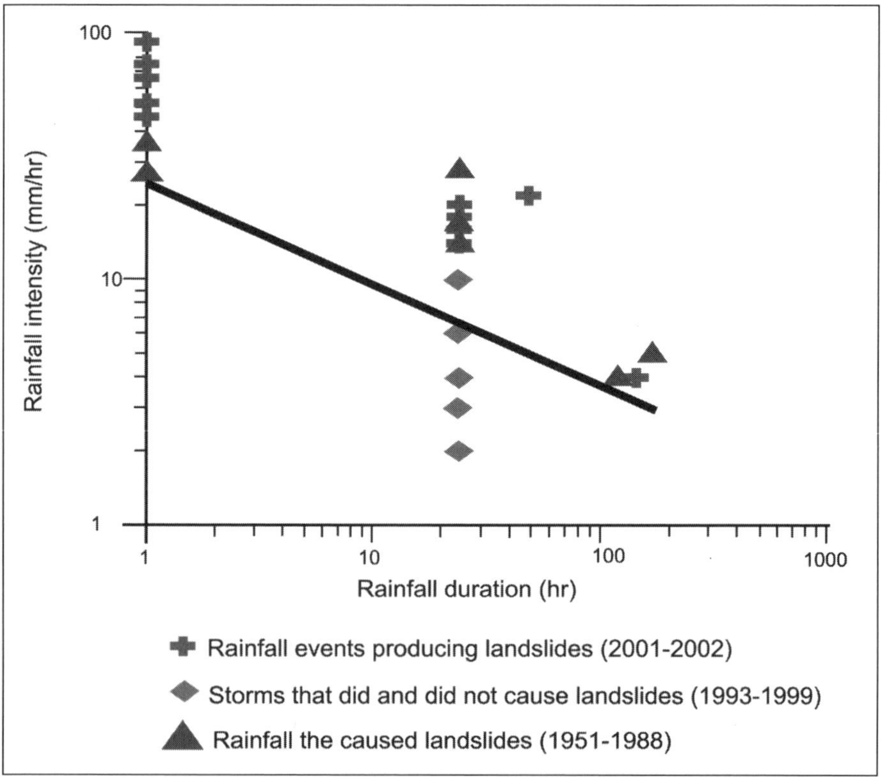

Figure 13.3 Threshold for rainfall-induced shallow landslides in Eastern Jamaica (Ahmad 2003).

Puerto Rico

Puerto Rico is a humid tropical island, where there is annual rainfall of approximately 2,000 mm. Landslides on this island have been triggered by long, intense rainfall. A study by Larsen and Simon (2006) conducted on 256 storms revealed that 41 of these events produced tens of thousands of landslides. The following formula was used to indicate the thresholds:

$I = 91.46\ D^{-0.82}$

I is the intensity in millimetres of the rainfall, and D is the duration in hours. It was also revealed that shallow slips and debris flows resulted from high-intensity rainfall events, and deep debris avalanches and slumps resulted from low-intensity, long-duration rainfall events (Larsen and Simon 2006).

Studies also revealed that for storms that had durations of up to 10 hours, landslides were only triggered when the rainfall intensity was as much as three times higher than rainfall intensity reported as sufficient to trigger landsliding in temperate regions. For durations of 100 hours,

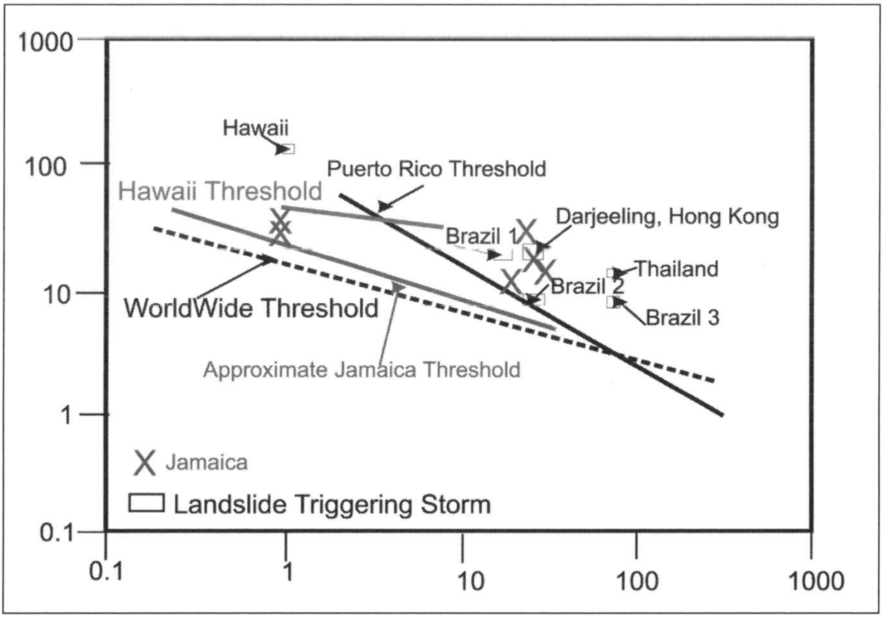

Figure 13.4 Rainfall intensities compared (Ahmad 2003).

the rainfall intensity necessary to produce landslides was that of the temperate regions (Figure 13.4). Furthermore, Jamaican data generally agrees with Caine's (1980) limiting curve. As storm durations approach 100 hours, all thresholds converge, indicating that geomorphological and climatological differences between humid tropics and temperate environments may not be significant when hillslopes receive large amounts of rainfall over a prolonged period. These indicate that the evolution of landforms is, to a significant degree, directly influenced by the landsliding process (Ahmad 2003).

Trinidad and Tobago

In July 2005, Tropical Storm Emily passed within close proximity of Trinidad. With winds of approximately 145 km/hr and heavy precipitation, the island was left with extensive flooding and multiple landslides along the northern coast. Daily rainfall in this area was approximately 117.2 mm. The features of 50 landslides (Figure 13.5) were investigated to determine the possible thresholds and the common characteristics of the affected hydrological catchments. Because the exact times of landslides were not recorded, the rainfall intensity/duration could not be determined. The landslides identified were located along roadways, as these were the only landslide locations recorded. The following list illustrates the most favoured characteristics of the catchments:

1. Soil type – sandy clay loam (82%)
2. Geology – slate, siltstone and sandstone (44%)
3. Slope angle – 20°–30° (32%)
4. Land use – forest (90%)
5. Elevation – 100–200 m (60%)

The combination of catchment conditions that were most favourable to landslides were as follows: forest (land cover); quartzite, phyllite (geology); sandy clay loam (soil); and 10–20 (slope) and 100–200 (elevation).

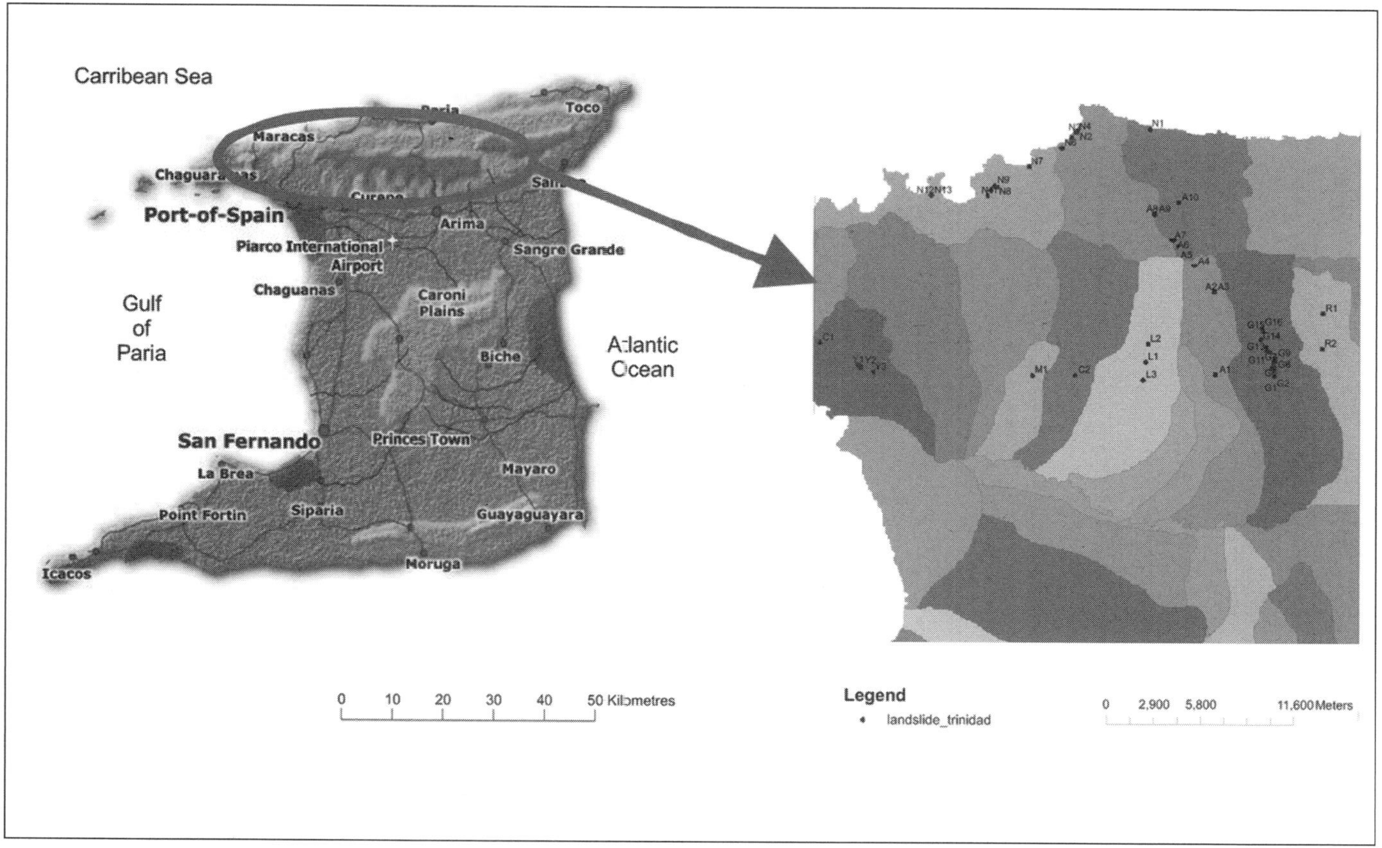

Figure 13.5 Location of 50 landslides that occurred during Tropical Storm Emily.

13.3 Developing an Early Warning System

Geoinformatics can be used to easily modify the parameters of models used that are the empirical intensity/duration models. Once the data has been collected – either remote or secondary data – the necessary layers can be analysed statistically. Table 13.2 indicates some of the functions that can be performed by the GIS that would be necessary for the operations of the early warning systems.

More specifically, the proposed early warning system for Tobago has been outlined in Figure 13.6. The development of this system would require the identification of the following:

1. Development of intensity/duration threshold: Research thus far has indicated that there is no inventory of dates and locations of landslides in Tobago. As a result, it is feasible to develop thresholds based on historical storm events, as done in Puerto Rico (Larsen and Simon 2006). Figure 13.7 (Pedrozzi 2004) displays the relationship between the rainfall intensity and duration and a stable

Table 13.2 Examples of the Use of GIS in the Early Warning System

Purpose	Layers
To determine the conditions favouring landslides and floods	Overlay of landslides and flood inventory map with soils map, DEM, rainfall data, geology, etc.
To determine where similar conditions occurs	Findings from above can be located to indicate areas that are also susceptible
Indication of relief sites	Themes indicating roads, hospitals, police stations, shelters and other critical resources
Maps indicating levels of warnings to be issued	Multi-layered buffers can be created from susceptible areas to determine the levels of warnings to be issued
Maps indicating the best evacuation routes	Overlay susceptible areas with roads, hospitals, police stations, shelters and other critical resources to determine the shortest evacuation path

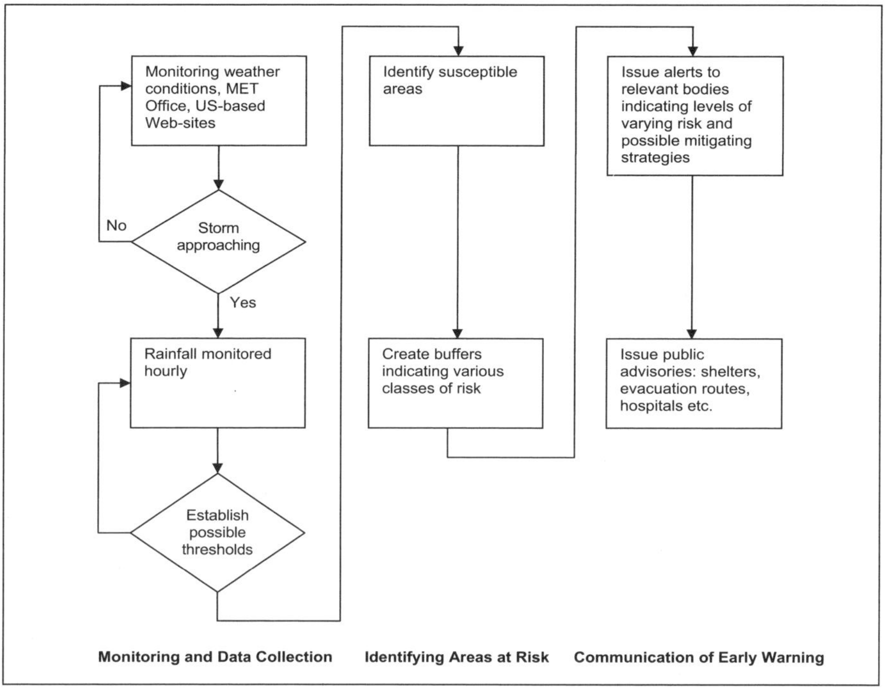

Figure 13.6 A proposed basic early warning system for Tobago.

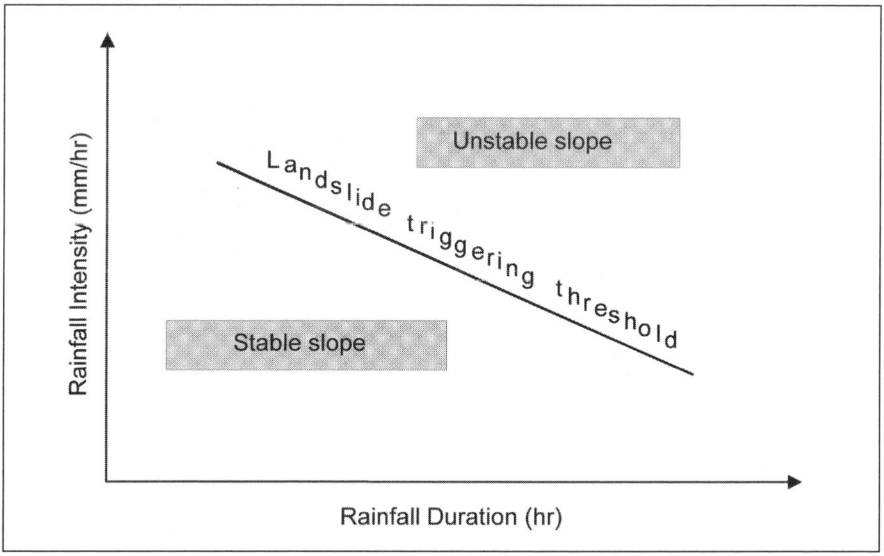

Figure 13.7 Correlation between rainfall intensity and duration, and the triggering of landslides.

and an unstable slope. Thresholds can be developed empirically through identifying historical storms, indicating which storms did or did not cause landslides and then linking this information to rainfall intensity/duration to derive landslide-triggering thresholds.

2. Identification of susceptible areas: These are the areas prone to the landslides if the thresholds are realized. A landslide susceptibility map, Figure 13.8, has been created for the island of Tobago (Baban and Sant 2004).
3. Location of critical resources: These include resources such as police stations, hospitals, shelters, evacuation route and the like, and they are used in determining the most efficient evacuation plans.
4. Communication of early warning signal: The public's perceived perception of the warning is vital, and issuers of warnings should appreciate that the psychological, community and individual processes in stressful times is more important than technology (ISDR 2004). It is important that people do not under- or overestimate the intended severity of the warning. One main proactive step

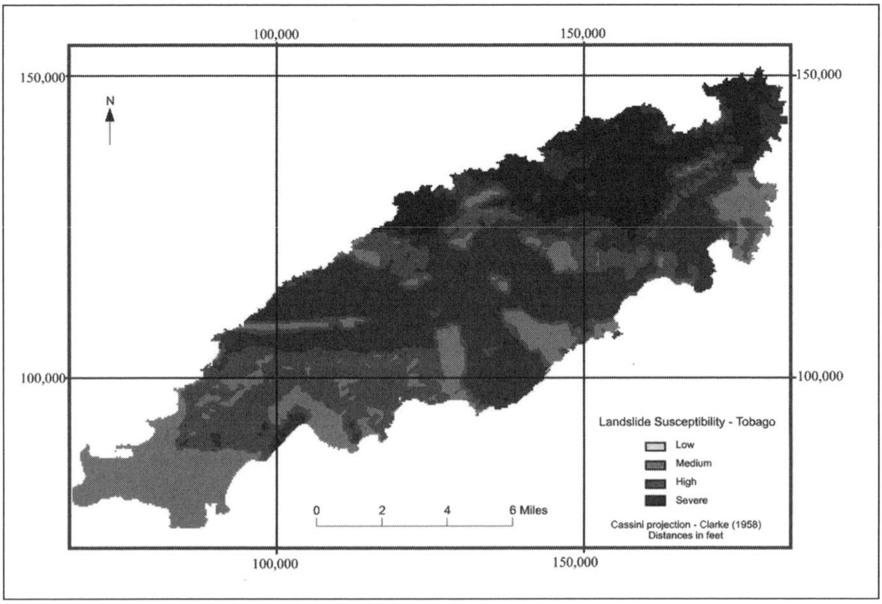

Figure 13.8 Tobago landslide susceptibility map (Baban and Sant 2004).

would be to involve the public in awareness and training programmes to ensure that they fully understand the severity of the warning, those who would be affected, the lag time and what is expected of them after the warning has been issued.

13.4 Conclusion

Landslides and floods are natural events, annual occurrences that can be hard to manage, and they can damage homes, livelihoods, critical resources and lives. In the tropical and mountainous Caribbean islands, these geohazards are rampant due to the abundance of suitable conditions. Structural and geological composition, together with increased deforestation and development activities, provide approving conditions for landslides and floods. This chapter has emphasized the need for an early warning system for managing these geohazards and the possibility of creating a geoinformatics-based early warning system for geohazards in the Caribbean. It showed how rainfall threshold values are determined and can be used as a basis for the system.

References

Aliasgar, K., and S.M.J. Baban. 2006. Developing a geoinformatics based early warning system for landslides in Tobago. *Proceedings of Urban and Regional Information Systems Association (URSIA) Conference*. The Bahamas.

Ahmad, R. 2003. Developing early warning systems in Jamaica: Rainfall thresholds for hydro-geological hazards. Paper presented at the Office of Disaster Preparedness and Emergency Management National Disaster Management Conference. Ochos Rios, Jamaica. http://www.mona.uwi.edu/uds/rainhazards_files/frame.htm.

Alexander D. 1995. *Natural disasters*. Dordrecht: Kluwer Academic Publishers.

Arenas D.A. 1983a. Hydrological characteristics of the Caribbean: Hydrology of the humid tropical regions with particular reference to the hydrological effects of agriculture and forestry practice. *International Association of Hydrological Sciences (IAHS)*, publication no. 140.

———. 1983b. Tropical storms of Central America and the Caribbean: Characteristic rainfall and forecasting. *International Association of Hydrological Sciences (IAHS)*, publication no. 140.

Baban, S.M.J., and K.J. Sant. 2004. Mapping landslide susceptibility on a small mountainous tropical island using GIS. *Asian Journal of Geoinformatics* 5, no. 1:33–42.

Caine, N. 1980. The rainfall intensity-duration control of shallow landslides and debris flows. *Geografiska Annaler,* ser. A, Physical Geography, 62:23–27.

Glade T., M. Crozier and P. Smith. 2000. Applying probability determination to refine landslide-triggering rainfall threshold using an empirical "antecedent daily rainfall model. *Pure Applied Physics* 157:1059–79.

Glantz, M.H. 2003. Usable science 8: Early warning systems – Do's and dont's. Report of workshop held 20–23 October 2003 in Shanghai, China. www.esig.ucar.edu/warning/

Gumbricht, T. 1996. Landscape interfaces and transparency to hydrological functions: Application of geographic information systems in hydrology and water resources management. *IAHS Publications* 235:115–21.

Helms D. 1989. *Flood warning systems: The basic elements. Proceedings: meeting of experts of hazard mapping in the Caribbean.* Kingston: University of the West Indies Publishers' Association.

ISDR. 2004. *Living with risk: A global review of disaster reduction initiatives.* Geneva: ISDR.

———. 2006. Platform for the promotion of early warning: In brief. http://www.unisdr.org/ppew/about-ppew/in-brief.htm. Last visited 9 April 2006.

Keller, R. 1983. Preface: Hydrology of the humid tropical regions with particular reference to the hydrological effects of agriculture and forestry practice. *International Association of Hydrological Sciences (IAHS)*, publication no. 140.

Larsen M.C., and A. Simon. 2006. Rainfall thresholds for landslides in humid tropical systems, Puerto Rico. http://www.pr.water.usgs.gov/public/webb/bibliography/abstract007.html.

LessLoss. 2005. Risk mitigation for earthquakes and landslides integrated project. Report on spatially distributed deterministic models and rainfall thresholds. http://www.lessloss.org.

Macpherson J. 1984. *Caribbean lands*. London: Longman Caribbean.
NOAA. 2005. Glossary. http://www.weather.gov/glossary/index.php?letter=f.
Pedrozzi, G. 2004. Triggering of landslides in the Canton Ticino (Switzerland) and prediction by the rainfall intensity and duration method. *Bulletin of Engineering Geology and the Environment* 63, no. 4:281–91.
Platform for the Promotion of Early Warning (PPEW). 2006. Basics of early warning. http://www.unisdr.org/ppew/whats-ew/basics-ew.htm.
United States Geological Survey (USGS). 2005. *The water cycle: Stream flow.* http://www.ga.water.usgs.gov/edu/watercyclestreamflow.html.

CHAPTER 14

Beyond Humanitarianism

Building Resilient Communities, Revisiting the Development Dialogue

JEREMY COLLYMORE

Abstract

The repeated and extensive loss to Caribbean society and economy from hazard impacts reflects a non-alignment between development planning practice and hazard management. Intervention in support of reducing hazard-related losses has traditionally been reactive and wanting. This chapter recognizes efforts to change this characterization and offers a framework for generating national readiness as a critical plank for promoting resilient development.

14.1 Introduction

The Caribbean region, however defined, has had a long history of disaster experiences associated with natural and human-influenced hazards. The impact on affected societies has been debilitating, often resulting in the retardation of planned development. In spite of this long history of disaster experiences, efforts are now only emerging to adopt and design development policies and practices informed by it

(Collymore 1992). Indeed, it may be suggested that it is this tendency to focus on hazard impacts more as events and less on the consequences that has resulted in an entrenched development practice in which hazard considerations are marginal at best.

The derailment of the economies in the region as a result of these impacts is well documented. The experience of CDERA states is indicative (Table 14.1).

Over the last three decades, disaster-related losses appear to be on the increase. Three times as many disasters were recorded in the 1990s as in the 1970s with similar increased levels in the cost of damage and in estimated persons affected (Rasmussen 2004; Freeman 2005).

A study of Eastern Caribbean states found that a natural disaster

Table 14.1 Value of Economic Losses from Disaster in CDERA Member Countries

Country	No. of Occurrences	Total Fatalities	Economic Losses ($m)	Economic Losses as % of GDP (1995)
Antigua and Barbuda	7	7	105.7	18.1%
Bahamas	4	5	290.4	9.5%
Barbados	5	3	148.4	6.3%
Belize	6	5	33.8	5.4%
Dominica	7	43	133.4	55.0%
Grenada	4	0	30.1	9.5%
Guyana	5	0	29.8	4.6%
Jamaica	19	271	1,988.1	29.3%
Montserrat	5	43	323.7	899.0%
St Kitts and Nevis	7	6	312.5	116.5%
St Lucia	8	54	1,554.6	272.3%
St Vincent and the Grenadines	9	5	47.0	16.5%
Trinidad and Tobago	8	9	16.7	0.3%

Source: IDB 2000; IDB, IMF, OAS, World Bank 2005.

occurs at least every 4.5 years in each one of them, affecting approximately 9% of the population and causing damage at 14% gross domestic product (Rasmussen 2004). The Jamaican experience is not too dissimilar (Tables 14.2 and 14.3).

Table 14.2 Selected Summary of Disasters Events: Jamaica, 1986–2005

Year	Natural Disaster	No. of Persons Killed	Total of Persons Affected	Total Damage in US$	GDP
1986	Flood	54	60,000	75,000	3.0%
1987	Flood				
1988	Flood	0	0	440	
1988	Hurricane Gilbert	49	810,000	1,000,000	65.0%
1991	Flood	0	551,340	30,000	6.0%
1993	Flood	9	4,372	11,000	
1994	Tropical Storm Gordon	4	0	0	
1995	Flood				
1996	Flood				
1996	Tropical Storm Marco	0	800	3,000	
1998	Flood			5,375	
2000	Drought	0	0	6,000	
2001	Flood				
2001	Hurricane Michelle	1	6,000	54,888	1.0%
2002	Flood	9	25,000	1,114,300	0.7%
2002	Hurricane Lili	4	1,500	0	
2002	Hurricane Isidore	0	0	0	
2004	Hurricane Charley	1	126	0	
2004	Hurricane Ivan	15	350,000	595,000	8.0%
2005	Hurricane Dennis	0	8,000	0	
2005	Hurricane Emily	4	2,296	0	
2005	Hurricane Wilma	1	100	0	
2005	Flood				

Source: CDERA 2005.

Table 14.3 Incidences of Floods in Jamaica

Year	Damage
June 1979	"New Market Flood"
August 1980	"Hurricane Allen – Agriculture (110m) hit hardest"
June 1986	June Floods
April 1987	"Floods destroy $167m in food crops"
November 1987	"$52 million road damage by floods, $73 million farm loss
September 1988	Hurricane Gilbert: massive islandwide destruction
June 1991	June: Floods damage to crops and livestock islandwide
January 1993	Floods: St Thomas suffers millions in damage
May 1993	Flood rains damage $400 million in agricultural crops
June 1993	Hurricane season starts: First tropical depression develops; rains pelt island
January 1993	"St Thomas flood damage runs into millions"
May 1993	"Millions needed for road repair" $25 million
November 1995	Island suffers flood damage
October 1996	169 families affected by floods
June 1997	June floods wreak havoc
December 1998	Flood leaves million damage
May 2001	Flood damage to crops especially in St Mary, St Ann
November 2001	Floods effects of Hurricane Michelle, Portland devastated
May 2002	Flood damage Clarendon, Manchester, St Thomas hardest hit

Source: CDERA 2005.

It may be accurate to suggest that the major setback to Jamaican and Caribbean economic development in the last three decades has been the frequent hazard impacts and the inadequate planning for them in the national economy (Pelling and Uitto 2002; IDB, IMF, OAS and World

Bank 2005). Indeed, what we have come to characterize as external shocks may in reality be inherent features of the economy.

Disaster management practice in our region has, however, been driven by a culture of crisis management. Our institutional arrangements are centred on anticipating the crisis and responding to it.

Traditionally, disaster management practice is humanitarian focused, though this is changing. Like many of development policies in the region, the emphasis has been on the management of symptoms. This, in many ways, explains why fatalities from disasters have decreased and socioeconomic losses continue to increase (Table 14.2).

14.2 Refocusing Disaster Management Interventions

It is therefore suggested that significant reduction in disaster losses will result only if disaster management is seen as a developmental problem, and it is appropriately integrated into the planning process. This perspective does not undervalue the benefits of knowledge enhancement or knowledge and technology transfer in reducing vulnerability. Indeed, it can be argued that the challenge is not one of awareness but one of embracing the available knowledge in policy and practice. With an average of forty disaster events a year, the Latin American and Caribbean region ranks second, only to Asia, in terms of frequency of disaster experience (IDB 2000). There are annual losses of more than US$3 billion per year. There is an indication that if a more comprehensive assessment of the impact of the event on the private sector is undertaken, the losses therein would be at least doubled.

While there has generally been good relief and response assistance by the regional and international community, this represents a small percentage of relief and recovery costs, very often less than 20%. The result is that significant budget re-adjustments and loan negotiations become necessary for affected states after a disaster impact. Even in those CDERA participating states, which have frequent disaster events, the strategizing for this resource gap is still primarily a reactive exercise.

A close analysis of the macro-economic assessment of recent disaster events will reveal that the fundamental problems of development that we face contribute significantly to the realized vulnerability (IDB 2000;

Rasmussen 2004). The observed vulnerability in the region is very much intertwined with rapid urbanization, entrenched poverty, environmental degradation, and limited policy coordination and implementation (UNISDR 2002; Comfort 1988).

Unfortunately, the burden in meeting these post-impact costs usually falls on the state. These increased demands come at a time when the state cannot afford to meet them. The initiation of relief initiatives without reference to coping capacity exacerbates the disconnectedness between disaster management and the realities of government financing.

Disasters serve as evidence of the need for change in public policy and practice and create opportunities to redesign, revise or rebuild the human environment damaged by the event (Comfort et al. 1999).

While it has been argued that small states' vulnerability be taken into consideration in programmes of assistance provided by the multi-lateral development finance and trade institutions, evidently vulnerability is not taken into considerations with respect to regional development and finance planning. It is this recognition of the inextricable link between disasters, development and the environment that informed a consultative dialogue among key regional stakeholders to fashion a framework for structuring this integrability.

14.3 The Regional Strategy

The goal of the Regional Strategy and Results Framework for Comprehensive Disaster Management (CDM) is sustainable development. Its strategic objective is to integrate disaster management into the development planning process (Figure 14.1).

Much of the promotion of this strategy has focused on the five intermediate results:

1. Stronger regional and national institutions promoting CDM
2. Research, education and training support CDM
3. Major regional institutions and donors incorporating CDM in their own programmes and promoting CDM to their national members/clients
4. Preparedness, response and mitigation capability is enhanced and integrated

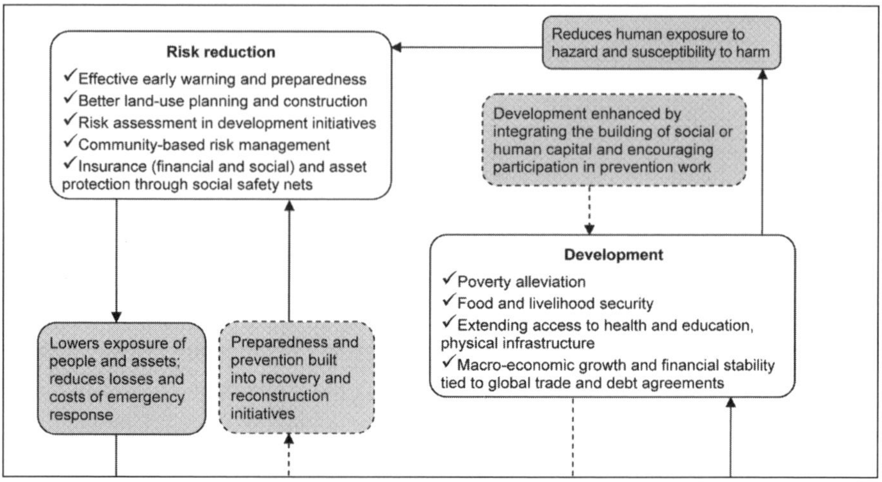

Figure 14.1 Regional strategy and results framework for CDM framework (CDERA 2005).

5. Hazard information is incorporated into development planning and decision making

The associated programme of work is aligned to the United Nations Hyogo Framework for Action on disaster risk reduction and is the platform around which such efforts are being engaged in the region, cooperative programming initiated and partnership management structured.

The region has not yet adequately addressed the values and principles that must inform the environment in which the intermediate results are fashioned. There needs to be a set of core values and principles that inform the phenomenology, which underpins the development process. Here, the issue of governance becomes central, as we seek to address the issues of organizational and distributional equity that will naturally surface in this realignment of the development agenda.

In the case of Jamaica, certainly there will be a need to examine the mechanism for best harnessing the efforts of the Planning Institute of Jamaica, the Urban Development Corporation, the National Environmental Planning Agency, the National Works Agency and others towards this goal. There is a need to examine the decision-making roles and levels of various actors, instruments used to engage discourse, mechanisms for fashioning contributions of stakeholders into policy programme and modalities of stakeholder participation.

14.4 Towards Resilient Development

Resilient development must be the platform on which public sector investment programmes are articulated. A resilient development agenda in this hazard-prone region forges an essential, but often avoided, dialogue on acceptability of risk. Such a dialogue will take us into the issues of desirability and context of insuring public facilities: who ought to be direct beneficiaries of relief policy; whether risk indifference should be compensated in relief policies; the nature and extent of disaster loss reduction incentives; and the generation of revenue in the emergency services.

Resilience puts emphasis on coping capacity. It takes away the emphasis from disaster losses and focuses it in a search for patterns of human organization and relationships that may reduce the cost of disasters (Pelling and Uitto 2002). What this should engender is a conscious trade-off between the costs and benefits of upfront and back-end investments in risk reduction or, more appropriately, the desirable mix of disaster loss reduction options.

Two critical steps are necessary to pursue a development dialogue anchored in the principles of resilience:

1. Formulation of a national disaster management policy: Much investment has been made in what may be characterized as operational policies. This is consistent with the techno-focus value neutral approach to problem solving that has engulfed our programme articulation.

   National disaster management policy is now required to provide the critical philosophy about the approach to risk management, including the values that will underpin the process.
2. Development of a national economy preparedness programme and plan: This will outline the steps in implementing the national disaster management policy. The objectives of such a programme would be
   a. to develop methods to reduce losses to critical elements of our communities;
   b. to develop methods to reduce the economic impact of disasters on individuals and families, small business, principal economic sectors, and the countries' overall economy;

c. to seek to determine the most appropriate uses and types of foreign aid in risk reduction and recovery financing;
d. to determine viable alternative economic activities that could reduce disaster loss potential;
e. to identify opportunities created by disasters for redirecting economic activities (for example, diversification and relocation); and
f. to utilize activities that would seek to identify opportunities created by disasters for expanding development programme activities.

The steps in this process will include the following:

1. Undertaking of hazard assessments
2. Estimation of vulnerability loss potential
3. Establishment of priorities for vulnerability reduction based on number two above
4. Preparation of an economic mitigation plan, which entails consideration of
   a. modification of development programmes to reduce loss potential;
   b. development of recovery programmes and institutional mechanisms;
   c. establishment of the financial and socioeconomic framework for relief policies; and
   d. development of community level risk management subprogrammes.

In essence, what is being proposed is a national economic disaster loss profile to assist physical, economic and social planners to structure loss reduction interventions, using one common information source. The national economic disaster loss profile provides the architecture for determining acceptable levels of risk, institutional roles and partnership arrangements, and geographic and sector priorities. With such a programme, the infrastructure will be developed for linking disaster loss reduction to Poverty Reduction Strategy, United Nations Development Assistance Frameworks and donor country assistance strategies. The

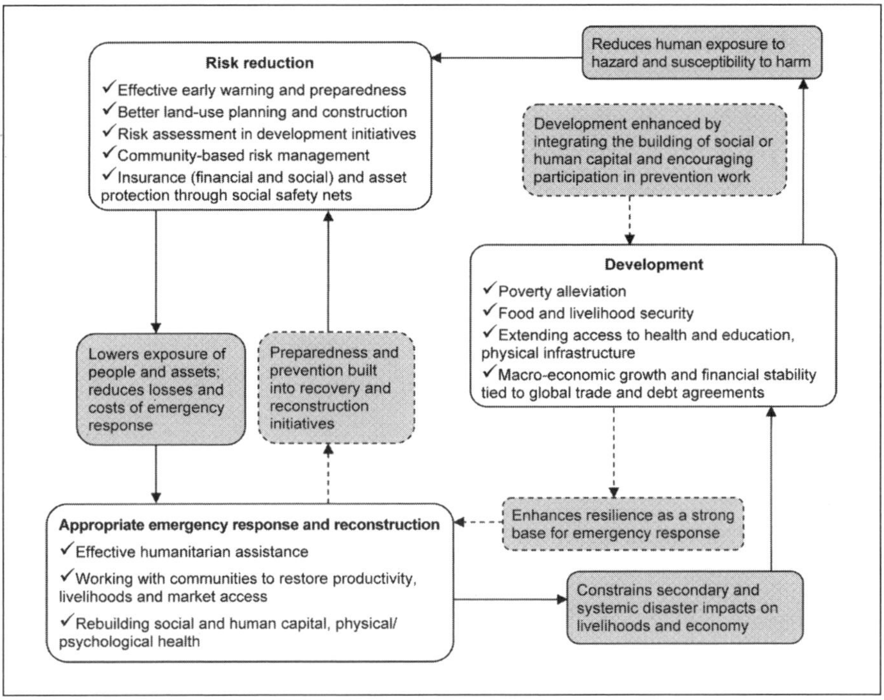

Figure 14.2 "Virtuous spirals" of risk reduction (DFID 2005).

outcome of this process may be what the Department for International Development (2005) calls the "virtuous spirals" of risk reduction (Figure 14.2).

14.5 Conclusions

It is clear that our development agenda is not directly addressing one of the greatest challenges to its sustainability, which is disaster-related loss, in spite of the unquestionable evidence of these increasing. The techno-solutions pursued are not underpinned by a development philosophy or practice that recognizes the threats of the operating environment risks. Progress in loss reduction, where achieved, is usually constrained to some aspect of a sector.

A development philosophy and approach centred on the generation of resilience, in all aspects of the society and economy, has been offered as a bridge for structuring a pragmatic agenda for reshaping the loss profile of our national economies.

References

Caribbean Disaster Emergency Response Agency (CDERA). 2005. Caribbean Community regional programme framework 2005–2015. Paper presented at the Second World Conference on Disaster Reduction. Kobe, Japan.

Collymore J. 1992. Planning to reduce the socio-economic impacts of natural hazards on Caribbean society. *Journal of the Geological Society of Jamaica* 12:88–97.

Comfort, L.K., ed. 1988. *Managing disasters: Strategies and policy perspectives. Durham: Duke Press Policy Studies.*

Comfort, L., et al. 1999. Reframing disaster policy: The global evolution of vulnerable communities. *Global Environmental Change* part B: *Environmental Hazards* 1, no. 1:39–44.

Department for International Development (DFID). 2005. Disaster risk reduction: A development concern. A scoping study on the links between disaster risk reduction, poverty and development. http://www.dfid.gov.uk/Pubs/files/drr-scoping-study.pdf.

Freeman, P. 2005. Background paper for thematic session on integrating disaster management into development financing: The role of international financial institutions. Paper presented at the World Conference on Disaster Reduction, Kobe, Japan.

IDB 2000. *Facing the challenges of natural disasters in the Latin America and the Caribbean: An IDB plan of action.* Washington, DC: IDB.

IDB, IMF, OAS, World Bank. 2005. *The economic of disaster mitigation in the Caribbean: Quantifying the benefits and costs of natural hazard losses; lessons learnt from 2004 hurricane seasons.* Working paper. Washington, DC: IDB.

Pelling, M., and J.I. Uitto. 2002. Small island developing states: Natural disaster vulnerability and global change. *Environmental Hazards* 3:49–62.

Rasmussen, T.N. 2004. *Macroeconomic of natural disasters in the Caribbean.* IMF Working Paper. Washington, DC: IMF.

UNISDR. 2002. *Living with risk: A global review of disaster reduction initiatives.* Geneva: United Nations.

Contributors

Serwan M.J. Baban is Professor of Environmental Geoinformatics, Head of the School of Environmental Science and Management, and Director of the Centre for Geoinformatics Research and Environmental Assessment Technology, Southern Cross University, Australia. He was formerly Professor of Surveying and Land Information, Chairman of the School for Graduate Studies, and Research Coordinator for the Centre for Caribbean Land and Environmental Appraisal Research, University of the West Indies, St Augustine, Trinidad and Tobago.

Rafi Ahmad is Head of the Unit for Disaster Studies, Department of Geography and Geology, and a Fellow of the Mona Geoinformatics Institute, University of the West Indies, Mona, Jamaica.

Kelly Aliasgar is an affiliate to the Centre for Geoinformatics Research and Environmental Assessment Technology and a doctoral candidate in the School of Environmental Science and Management, Southern Cross University, Australia.

Raid Al-Thair is Senior Lecturer in Surveying and Land Information and Coordinator of the Centre for Caribbean Land and Environmental Appraisal Research, University of the West Indies, St Augustine, Trinidad and Tobago.

Francis Cannisus is a postdoctoral scientist affiliated with the Centre for Caribbean Land and Environmental Appraisal Research, University of the West Indies, St Augustine, Trinidad and Tobago.

Jeremy Collymore is Coordinator of the Caribbean Disaster Emergency Response Agency, Barbados.

Angela Cropper is Assistant Secretary-General and Deputy Executive Director for the United Nations Environment Programme in Nairobi, Kenya.

Andrew Fox is Senior Lecturer in Civil Engineering Management, Faculty of Engineering and Computing, Coventry University, United Kindgom.

Ronnie Kantasingh is a professional surveyor, an affiliate with the Centre for Caribbean Land and Environmental Appraisal Research, University of the West Indies, St Augustine, Trinidad and Tobago.

Bheshem Ramlal is Lecturer in Surveying and GIS, and a member of the Centre for Caribbean Land and Environmental Appraisal Research, University of the West Indies, St Augustine, Trinidad and Tobago.

John Ritter is Assistant Professor of Geology and Director of Environmental Studies Program, Wittenberg University, United States.

Kamal Sant is a professional surveyor, an affiliate with the Centre for Caribbean Land and Environmental Appraisal Research, and a doctoral candidate in the Department of Surveying and Land Information, University of the West Indies, St Augustine, Trinidad and Tobago.

Vernon Singhroy is Senior Research Scientist, Canada Centre for Remote Sensing.

Deborah Thomas is Assistant Coordinator, Town and Country Planning Division, Ministry of Planning and Development, Trinidad and Tobago.

Keith Tovey is Reader in Environmental Sciences, School of Environmental Sciences, University of East Anglia.